ACHTUNG-PANZER!

ACHTUNG-PANZER!

THE DEVELOPMENT OF TANK WARFARE

HEINZ GUDERIAN

Translated by Christopher Duffy

Introduction and Notes by Paul Harris

CASSELL

A CASSELL MILITARY PAPERBACK

First published in German in 1937
First English translation published by Arms & Armour 1992
This paperback edition published in 1999
by Cassell,
an imprint of Orion Books Ltd,
Orion House, 5 Upper Saint Martin's Lane,
London WC2H 9EA

An Hachette UK company

17

A CIP catalogue record for this book
is available from the British Library.

ISBN 978-0-3043-5285-2

Designed and edited by DAG Publications Ltd
Designed by David Gibbons; edited by Michael Boxhall

Printed and bound by
CPI Group (UK) Ltd, Croydon, CR0 4YY

The Orion Publishing Group's policy is to use papers that
are natural, renewable and recyclable products and
made from wood grown in sustainable forests. The logging
and manufacturing processes are expected to conform to
the environmental regulations of the country of origin.

www.orionbooks.co.uk

CONTENTS

EDITORS' INTRODUCTION

Achtung – Panzer! is one of most significant military books of the twentieth century. Guderian distilled into it about fifteen years' study of the development of mechanized warfare from its origins in the First World War until 1937, the year the book was completed and published. He sought to demonstrate that only by the intelligent use of armoured formations could Germany achieve swift and decisive victories in future wars, and avoid the ruinous attrition experienced in 1914-18. Although a number of conservative senior officers were sceptical of Guderian's message, by the outbreak of the Second World War it had gained a good deal of acceptance. The panzer (armoured) divisions became the cutting edge of the German Army in its spectacular victories of 1939-42.

Guderian was an outstanding soldier. A pioneer in the development of the German armoured forces, he later proved himself a dynamic and effective field commander, playing crucial roles in the Polish campaign, in the operation against France in May-June 1940 and in the attack on the Soviet Union in 1941. He was removed from field command in December 1941 after disputes with his superiors and for about fifteen months was unemployed before becoming Inspector of Panzer Troops in March 1943. In July 1944 he became Chief of the General Staff but was sent on indefinite leave in March 1945 after disagreements with Hitler.[1] Although he was interrogated about war crimes no serious evidence was found against him and he was released without being indicted. He died in 1954.[2]

An English edition of Guderian's memoirs, entitled *Panzer Leader*, was published in 1952 and is one of the best known German accounts of the Second World War. Yet *Achtung – Panzer!*, which explains the thinking behind the operations of the panzer forces and which served as a textbook for trainee panzer officers during the war, seems never to have been fully translated into English and has not been widely read outside Germany.

Heinz Guderian was born in Kulm on the River Vistula in East Prussia in 1888. Military service was not a longstanding family tradition, but Heinz's father, Friedrich, himself an officer, decided on a military career for Heinz when the latter was still quite young. Guderian's secondary education was, therefore, in cadet schools specifically designed to prepare boys for entry to the officer corps. Guderian looked back with affection on his school years. Although the atmosphere was disciplined and conditions spartan he remembered his teachers as humane. The academic curriculum was very similar to that of the Realgymnasia – the leading German civilian secondary schools – consisting largely of modern languages, maths and history. The standards were, Guderian insisted, equally high. A deep respect for education was a vital part of the Prusso-German military tradition of which Guderian was an outstanding product.

Particularly significant for the future development of Guderian's military thought were the linguistic abilities he began to acquire at school. He developed excellent French and good English. Knowledge of these languages together with his native German opened up to him much of the military literature of Europe. *Achtung – Panzer!* is an eclectic book in which Guderian selects with great discernment and synthesises with great intelligence ideas drawn from a considerable range of sources.[3]

In February 1907 Guderian was sent as an ensign-cadet (a trainee-officer not yet commissioned) to the unit commanded by his father – the 10th Hanoverian Jäger Battalion. Jäger troops in the German Army were the equivalent of the light infantry or Rifle Brigade troops in the British Army and their traditions encouraged rather more personal enterprise and initiative than those of ordinary infantry. After attending the War School at Metz from April to December 1907 he was commissioned into his battalion as a Second Lieutenant in January 1908. He married in 1913.[4]

Guderian was attached to a telegraph unit from October 1912 until September 1913 and served with signals for much of the First World War. During that war he seems to have seen very little front-line service. He became a radio-specialist and later a staff officer. An accurate understanding of the military possibilities of radio was very important for the development of Guderian's military thought.[5]

His staff training was even more significant for his future career. Guderian was proud to be a member of the German General Staff – the intellectual élite of the Army which formulated its doctrine and made its war plans. His memoirs are largely about his experience in the Second World War and tell us relatively little about his General Staff education and training, but he does record that he was attached to the War Academy in Berlin (which was the General Staff school) from October 1913 until the outbreak of war. His General Staff training was interrupted by the events of August 1914 and not formally completed until he attended a short course early in 1918.[6]

Achtung – Panzer! is in many ways a typical product of the thought processes developed by a German General Staff education. A thorough grounding in German military history and the inculcation of the habit of detailed analysis of recent military operations in order to draw out lessons for the future were among its most important features. Contrary to a common opinion in the English-speaking countries, German military training at this highest level was not intended to produce robots each programmed to think in exactly the same way. It was designed to develop and disseminate a common doctrine which all officers could understand and implement. But students were also taught that doctrine must evolve in accordance with political and technical change.[7] If they qualified as members of the General Staff they were expected to play a part in this evolutionary process. That, of course, was Guderian's purpose in *Achtung – Panzer!*.

For a regular officer who had served throughout a world war Guderian's career progression had been remarkably slow – he was still a captain in 1919. But his signals training and his General Staff education were to stand him in good stead. The Treaty of Versailles, imposed on Germany by the Allies in 1919, theoretically abolished the German General Staff as well as banning the Germans from having tanks, submarines and many other types of weapon. But the fact that Germany's enemies wanted to destroy it appears merely to have emphasized the importance of the General Staff in the German military mind. (It continued to exist under another name, being called the *Truppenamt* (Troop Office) for most of the inter-war period.) One of the most significant years in his entire career was 1922 when General Tschischwitz, the head of the Motor Transport Troops, asked for a General Staff educated officer to help him consider the possible application of motor transport to a combat role.[8]

Tschischwitz was empire building. The transport of food and ammunition was an important activity but hardly glamorous. Tschischwitz was seeking a more prestigious role for the forces he commanded. Germany still faced her usual problem of having potential enemies on widely separated fronts. It was possible that she might become involved in a war with France and Poland or with France and Czechoslovakia. In the 1920s the size of the German Army was fixed by the Treaty of Versailles at a mere 100,000 men. It could not hope to gain a decisive victory in any such conflict. But by making the best use of its limited strength it might stave off decisive defeat until the rest of the international community intervened to stop the war, perhaps under the aegis of the League of Nations. One way of doing this might be to employ a mobile operational reserve which could strike a hard blow on one front and then move rapidly to mount a counter-attack on the other. Motor transport would confer much greater flexibility on such a reserve than railways ever could.[9]

Having carefully studied the concept of motorized combat troops, Guderian came to the conclusion that it was not merely valid but vital for the future of the army. Infantry mounted in trucks, however, would not be sufficient in themselves. They would need to be combined in fully motorized formations with the traditional supporting arms – artillery and engineers – and also with tanks.[10]

Although during the 1920s he became fascinated with them, Guderian had had no involvement with tanks in the First World War. Very few German officers had. Germany was slow to introduce tanks, not even starting to develop them until they had been used by the British on the Somme in September 1916, and giving them a very low priority even when the British had demonstrated their potential at the Battle of Cambrai in November 1917. Even at the end of the war so few German tanks had been manufactured that they were outnumbered by captured Allied tanks in front-line German units. Because tanks had been banned by the Treaty of Versailles, the Germans in the mid-twenties had only a few experimental models and these were being tested in the Soviet Union (by agreement with the Soviet government) in conditions of considerable secrecy.[11]

First-hand experience was therefore rather difficult to come by. But Guderian did not allow this to get in his way. He made personal contact with some of the relative handful of German tank veterans from the First World War and read everything he could lay his hands on about tanks. By 1937 his reading appears to have included British works by Swinton, Fuller and Martel and a short book by an obscure French officer – one Charles de Gaulle. It says something for the broadmindedness of the German Army in the 1920s that Guderian was officially acknowledged as a leading expert on tank tactics years before he first set foot in a tank.[12]

Already in the twenties there were tanks available in foreign countries, notably the British Vickers Medium, which were fast enough and had long enough ranges to be considered not merely of tactical but potentially of operational significance. Guderian came to the conclusion that when tanks were incorporated in mechanized formations with infantry, artillery and engineers in correct proportion, when several of these formations were available, and when they were used together for concentrated blows, they might determine the course of campaigns. This was Guderian's crucial insight and he claims to have made it by the beginning of the 1930s.[13] From then on until the outbreak of the Second World War it was with the creation of motorized combat units and formations and with the development of a doctrine for their use that Guderian was principally concerned.

Teaching and writing absorbed a good deal of Guderian's time in the 1920s. Teaching military history and tactics to transport corps officers destined for staff work was Guderian's main official duty in the period from October 1924 to October 1927. The officers he taught proved stimulating and demanding audiences. Guderian seems to have gained a considerable reputation as a teacher and when he was transferred to the transport department of the War Ministry in October 1927 he was concurrently employed as a lecturer on tank tactics to the Motor Transport Instructional Staff, an activity which he seems to have kept up until 1930. From 1924 to 1935 he also contributed many articles to the military press especially the *Militär-Wochenblatt* (Military Weekly) whose editor, General Altrock, gave him a great deal of encouragement. Writing and teaching, together with wargaming, seem to have played an important part not only in the dissemination but also in the evolution of Guderian's military thought.[14]

In 1929 Guderian encountered real tanks for the first time. On a trip to Sweden he was given hospitality by a Swedish tank battalion which permitted him to learn to drive one of their M21 tanks. These machines were a version of the LK II which had been developed by the German Army in the First World War but had not seen active service before the war ended. Some of them had then been sold to Sweden. The M21 was a not particularly well-designed vehicle and by 1929 it was obsolete. Nevertheless Guderian records that he found his Swedish visit instructive.[15] Having at last gained some 'hands on' experience, however limited, must have boosted his confidence and enhanced the conviction with which he taught and wrote about tank operations.

In February 1930 Guderian, at the suggestion of his friend Colonel Lutz who had become Chief of Staff to the Inspectorate of Transport Troops, took command of the 3rd (Prussian) Motor Transport Battalion. This battalion was being trained as a motorized combat unit and was equipped with some real armoured cars and motor-cycles and with dummy tanks and anti-tank guns. The unit was involved in occasional exercises in which it attempted to demonstrate the utility of mechanized units to the rest of the army, but generally Guderian received little encouragement from his superior, General Otto von Stülpnagel, the Inspector of the Transport Troops. According to Guderian this officer expressed the opinion that 'neither of us will ever see German tanks in operation in our lifetime'.[16]

Fortunately, from Guderian's point of view, Stülpnagel retired in 1931, and even more fortunately he was succeeded by Lutz. In October 1931 Guderian became Lutz's chief of staff. The Lutz–Guderian partnership seems to have been absolutely vital to the development of the German mechanized forces in the critical period up to the establishment of the first three panzer divisions in October 1935. Although Lutz was the more senior officer, Guderian is usually regarded as having been the intellectual driving force.[17]

One area in which Guderian's influence proved to be of considerable importance was communications – the key to command and control. Enormous effort was dedicated to providing tanks with effective radio sets. By the outbreak of war every German tank had at least a radio receiver and every command tank had a transmitter also. This was a much better provision than that in any other army at this time and it gave the German armoured formations an extremely high degree of tactical flexibility.[18]

By the time Hitler became Chancellor, in January 1933, the military concepts on which the *Panzertruppen* (armoured forces) were based had been quite clearly formulated by Guderian and Lutz. There was nothing distinctively Nazi, therefore, about the panzer divisions or the style of warfare which they helped to make possible. Yet Germany had a mere handful of real tanks when Hitler came to power and most training was done with canvas dummies. The extraordinary growth of the panzer arm from then until 1940 was owing to the overriding priority which Hitler gave to Germany's rearmament, helped to some degree by his personal enthusiasm for modern technology and particularly for the internal combustion engine.[19]

This enthusiasm is well illustrated by an incident which took place at the tank proving grounds at Kummersdorf in (according to most authorities) early 1934.[20] After witnessing a demonstration of motorized troops, including most of the basic elements which were later to compose the panzer divisions, Hitler is reported to have exclaimed, 'That's what I need! That's what I want to have.' Yet the panzer forces never gained an overriding priority. Lutz and Guderian were forced to compete fiercely for resources with the other arms. The panzer divisions formed only a small proportion of the German Army when the Second World War began and they remained so throughout its course.[21]

From October 1931 to October 1935 Guderian, as Lutz's chief of staff, had been at the very heart of affairs, first at the Inspectorate of Motorized Troops in the Defence Ministry and, from 1 July 1934, at the new Armoured Troops Command. When the first three panzer divisions were established, however, Guderian was given command of 2nd Panzer Division based at Würzburg, an appointment which physically removed him from the centre of policy-making. (He was replaced as Lutz's chief of staff by Colonel Paulus who, as a Field Marshal, was later to be captured by the Russians at Stalingrad.) Lutz remained Guderian's immediate superior, however, and it was on Lutz's instructions that, in the winter of 1936-7, he wrote *Achtung – Panzer!*.[22]

At that time Lutz and Guderian seem to have believed that there were still major intellectual and institutional battles to be fought before the panzer troops could be assured of their rightful share of the resources devoted to Germany's rearmament. While some of the most senior officers in the army were generally supportive of the work that Lutz and Guderian were doing there was still considerable scepticism among others. Guderian regarded Colonel-General Beck, Chief of the General Staff 1933-8, as a major obstacle to progress. Beck was of a somewhat cautious and conservative disposition and, though by no means opposed to the use of tanks, had somewhat different views from Guderian on their distribution and employment.[23]

One development of the second half of the 1930s which particularly worried Guderian (and which was partly attributable to Beck) was the raising of tank brigades for close co-operation with infantry divisions. Another was the formation, at the instigation of the cavalry, of so-called Light Divisions – mechanized divisions consisting largely of cavalry personnel which were primarily designed for reconnaissance and screening, the main cavalry roles of the late 19th century. The motorization of four infantry divisions also dismayed him. These developments seemed to Guderian to dissipate, in a potentially ruinous way, the German Army's assets in tanks and other motor vehicles. Given how very limited these resources were, Guderian believed that they should be concentrated almost exclusively in the panzer divisions. He thought the infantry divisions should continue to move on foot and rely on horses to draw their transport and artillery. The Light Divisions, while tying up some tanks and a lot of motor transport, lacked offensive power and would be of little use. The employment of tanks merely in support of unmechanized infantry was a reversion to the practice of 1916-18. Tanks in this role might have some limited tactical utility but could have no operational impact.[24]

Guderian informs us that Lutz instructed him to write *Achtung – Panzer!* in order to gain the widest possible publicity for the cause of the panzer divisions.[25] It is thus very much a tract for the times, intended to score points off institutional opponents and to gain the maximum resources for Guderian's own branch of the Army. Given this polemical purpose, and given that Guderian wrote it in just a few months while commanding a division, the reader may well be surprised at how serious and substantial a treatise *Achtung – Panzer!* turned out to be. By 1936 Guderian had been immersed in the study of mechanized warfare for so long that it was probably all but impossible for him to produce a glib piece of mere propaganda even if that were all Lutz required of him.

Achtung – Panzer! is a work of theory which was intended to help Germany prepare for the warfare of the immediate future. But it is also a work of history. More than half the book is devoted to a rigorous analysis of the experience of the Western Front in the First World War. The emphasis is on the factors which brought the tank into existence, the technical development of tanks, the organizational development of tank corps, and the actual experience of tank operations.

Guderian's historical analysis is remarkable for its lack of national chauvinism. Inevitably he praises the British and French for being the front-runners in tank development and deprecates Germany's tardiness in producing tanks of her own. But his frankness goes beyond that. It had suited the Nazis to claim that the German Army had not been defeated in the field but stabbed in the back by Jews and Socialists.[26] But as part of his advocacy of the importance of the tank, Guderian openly admits that the German Army was conclusively defeated on the Western Front – the one theatre of operations in which tanks were employed in substantial numbers. It assists his thesis to point out that the negotiations leading to the Armistice were sought, on the insistence of Ludendorff, the Army's supreme commander, as a direct result of the shock administered at Amiens on 8 August 1918, by a British offensive led by 400 tanks, and he makes this point too without equivocation.

Guderian's account of military developments in the First World War and of the role of the tank in that war can be criticized in some particulars. Many of the works he consulted were written by actors in the drama and cannot be regarded as objective. Using Major-General Swinton's memoirs, *Eyewitness*, as one of his main sources, it is arguable that he somewhat over-rates Swinton's part in the development of the tank.[27] Perhaps because of the hurry in which he worked, his account is occasionally slightly garbled. He gives Winston Churchill credit for the suggestion of using the Holt caterpillar tractor as the basis for a trench-crossing machine, whereas Churchill, for all his drive and imagination, lacked the technical knowledge to make this suggestion[28] and in fact the Holt system played virtually no part in the development of British tanks. Guderian relies largely on authors who were leading tank advocates and most present-day military historians would argue that the capacity of the tank to influence events on the Western

Front was much more marginal, even in 1918, than Guderian suggests.

But these points matter little. Guderian was not writing primarily for an academic readership. What was significant to him was to understand the events of the recent past accurately enough to draw the correct military lessons for the future. Guderian was right in that as a force in European war cavalry was in terminal decline. He was also right to emphasize the enormous advantages that lay with the defender and the acute weakness of infantry in the offensive, a weakness that even massive artillery support had proved inadequate to offset. Perhaps Guderian does not give the German infantry sufficient credit for the great tactical strides which they made between 1914 and 1918. (Storm troops and the infiltration tactics which played such a crucial role in the initially successful German offensives of spring 1918 are hinted at but not actually mentioned.) Yet he was essentially correct that without mechanical assistance even the German infantry of 1918 could achieve breakthrough only at very great cost. It had proved quite incapable of the rapid exploitation necessary to convert mere tactical victories into decisive operational success.

From his detailed accounts of all the main tank actions of the First World War Guderian explicitly draws the following lessons:

(a) tanks are of little use when penny-packeted and should be massed;

(b) they should not be wasted on unsuitable ground, as they were by the British GHQ in the swamps of Ypres, but saved for use on good going;

(c) that the greatest results can be achieved when massed tanks are used with the benefit of surprise.

These were lessons which Ernest Swinton, J. F. C. Fuller and Giffard le Quesne Martel had been trying to drive home in the British Army from 1916 onwards. They were by no means original to Guderian. But the ability to learn from enemies is a military virtue and the lessons were no less valid for being acquired from the writings of British officers.

Although he does not devote a very great deal of space to it, Guderian is also extremely perceptive in his discussion of the role of air power. He points out that the importance of military aviation greatly increased during the course of the war and that, by 1918, the Germans were operating in the face of a substantial Allied air superiority. He is not impressed with the effectiveness of Allied air attack on the German homeland but stresses the impact of aircraft on operations on the Western Front. In what may be regarded as one of the most significant paragraphs in the book, he explains that Allied aircraft in 1918 created disorder in the German rear areas, hindered the movement of reserves and brought German batteries under actual attack.

'All of this was of material influence on the course of the ground fighting, especially when they were acting in co-ordination with tanks. Aircraft became an offensive weapon of the first order, distinguished by their great speed, range and effect on target.'

In the second half of the book Guderian is concerned with post-war military developments, especially in armoured fighting vehicle design and in the organization of mechanized formations. Even here his approach is largely historical. He first records what has already happened and only then prescribes for the future. It is all very pragmatic. There is no abstract theorizing. The developments Guderian suggests in military organization and in the conduct of operations are no more than logical projections of trends he had observed in the First World War, making allowance for technological change.

Guderian argues that it is ridiculous to waste the potential afforded by the much greater speeds of which tanks have become capable since the First World War by tying them to the pace of the infantry divisions. Tanks should, therefore, be concentrated in formations designed to exploit their potential – formations capable of both breakthrough and exploitation. Yet he does not argue that tanks can achieve much on their own. Armoured formations, he argues, must be large and must include other arms including infantry and artillery. The ideal solution, he indicates, is for the artillery to be self-propelled and for the infantry to be mounted in armoured carriers with roughly the same mobility characteristics as the tank. In fact, at the time he was writing, the panzer divisions' artillery was largely towed and there were no armoured carriers for their infantry component. As Guderian explains, the infantry of the panzer divisions were having to make do with motor-cycles and trucks for mobility.

The decision to commit the armoured divisions, he maintains, must lie with the high command. He insists that they should always be deployed in both breadth and depth and used *en masse*. In a particularly perceptive and telling sentence he argues that: 'Concentration of the available armoured forces will always be more effective than dispersing them, irrespective of whether we are talking about a defensive or offensive posture, a breakthrough or an envelopment, a pursuit or a counter-attack.' Guderian regarded the breakthrough battle as presenting great difficulties. The First World War had shown the inherent strength of the defence in an age of magazine rifles, machine-guns and quick-firing artillery. To smash through a well-prepared defence Guderian believed that it would be necessary to bring the whole depth of the enemy position under attack simultaneously. In this context he argues that air power will have a crucial part to play in support of the panzer divisions.

'Success is probably attainable only when the entire defensive system can be brought under attack at more or less the same time. When the attack begins the enemy hinterland must be subjected to vigilant aerial surveillance, so as to identify the movements of the enemy reserves and direct our combat aircraft against them. The air forces must bend their efforts to preventing or at least delaying the flow of those reserves to the location of the breakthrough.'

There are strong echoes here of the 'Deep Battle' ideas of Soviet military thinkers such as Triandafillov and Tukhachevsky. Indeed in the section of the book dealing with the development of armoured forces in Russia

Guderian shows some familiarity with Deep Battle theory. He mentions neither Tukhachevsky nor Triandafillov but does refer to and quote Kryshanovsky – a lesser known member of the same school.

Readers will notice that nowhere does Guderian use the term '*Blitzkrieg*' – often thought to encapsulate the German approach to war in the Nazi era. In fact the term seems not to have been used in Germany before the Second World War when it was picked up from the foreign press. Its first known use occurs in a 1939 article on the Polish campaign in the American *Time* magazine.[29] It conjours up rather well the image of a fast-moving campaign but is no more than a journalist's buzz-word, devoid of precise meaning. It never became part of the official German military vocabulary.

For a British historian one of the most striking things about *Achtung – Panzer!* is Guderian's devotion of so much space to British military ideas and innovations. Guderian was, after all, a member of the German General Staff – one of the greatest institutions ever created for solving the problems of large-scale land warfare. It is remarkable, therefore, that he draws so much inspiration from an island nation, generally much less adapted to this type of conflict than the Germans.

There is, however, no doubt that up to the end of the 1920s (while tanks were banned in Germany) the British led the world not only in the technical development of fighting vehicles but also in the organization and handling of armoured formations. Guderian was obviously fascinated by the British General Staff's pioneering work with the Experimental Mechanical Force, established in 1927. He also took an interest in the experimental Tank Brigade of 1931 and the Mobile Force exercise of 1934. The British appeared to Guderian to be much more interested than the French in the operations of mechanized formations deep in the enemy's rear. He freely acknowledges that, up to the early thirties, the Germans were using a translation of a British General Staff booklet as their own manual on mechanized warfare. Sadly, by the end of the decade the British had fallen badly behind the Germans in this field.[30]

On the subject of British influence, some readers will be surprised that there is not more reference to Basil Liddell Hart, the famous military journalist and author. He is mentioned only once – in connection with the Experimental Mechanical Force of 1927. Since the late 1970s military historians specializing in this period have been aware that the well-known passage in Guderian's memoirs, in which Liddell Hart is extolled above all others as the inspiration behind the early victories of the panzer forces, was put there at Liddell Hart's own request at a time when Guderian was in various ways indebted to him. Significantly this passage does not occur in the original German edition.[31] While British books figure prominently in the bibliography of *Achtung – Panzer!*, there is none among them by Liddell Hart and nothing in the text to indicate that Liddell Hart had any marked influence on the development of Guderian's thought.

The part of *Achtung – Panzer!* dealing with Russia now seems particularly prophetic. Guderian credited the Russians with having 10,000 tanks and was obviously impressed with the seriousness with which they approached the business of mechanized warfare. He finishes his discussion of the Soviet Union with a terrible warning:

'Russia possesses the strongest army in the world both in numerical terms and in terms of the modernity of its weapons and equipment. The Russians have the world's largest air force as well . . . Russia has ample raw materials and a mighty armaments industry has been set up in the depths of that vast empire. The time has passed when the Russians had no feeling for technology; we will have to reckon on the Russians being able to master and build their own machines and with the fact that such a transformation in the Russians' fundamental mentality presents us with the Eastern Question in a form more serious than ever before in history.'

Confronted with the evidence of this passage we probably ought to believe Guderian when he tells us that he was horrified at the prospect of invading the Soviet Union and wrote a memorandum to the high command opposing Operation 'Barbarossa'.[32]

On a personal level *Achtung – Panzer!* was a great success for Guderian. The book became a best-seller, the proceeds from which bought him his first car. It is impossible to tell to what extent its publication assisted the panzer forces in the battle for resources within the German Army. Probably its impact was only marginal. Yet the panzer forces developed impressively in the late thirties. Only three panzer divisions were established in October 1935, but by the outbreak of war there were six. In the Polish campaign the 'Light' divisions, established at the instigation of the cavalry lobby, were found (as Guderian had predicted) to lack sufficient offensive power and all four were converted to panzer divisions. Thus there were ten panzer divisions available for the attack on France in May–June 1940, their most dramatic victory.[33]

As they approach the end of their task the editors are more than ever convinced of the importance of Guderian as a military thinker. It is probably not going too far to say that *Achtung – Panzer!* is essential reading for anyone who wishes to understand the development of land warfare in the 20th century. The appearance of an English edition is surely long over-due.

Notes

1. H. Guderian, *Panzer Leader* (Arrow) 1990, passim.

2. K. Macksey, *Guderian: Panzer General* (Macdonald and Jane's), 1975, p. 208.

3. On Guderian's schooling see Guderian, op. cit., p. 16. On his linguistic abilities see Macksey, op. cit., pp. 7 and 205. On his eclecticism Brian Bond notes that he followed the development of new military ideas abroad with the aid of the publication *Wehrgedanken des Auslands* and that he set up his own translation service for English military books, articles and doctrinal pamphlets. B. Bond, *Liddell Hart, A Study of his Military Thought* (Cassell), pp. 221-2.

4. Ibid., p. 17.

5. Macksey, op. cit., pp. 9-10 and 50-2.

6. Guderian, op. cit., pp. 17 and 467-8.

7. A very useful summary of the rise of the German General Staff under the elder Moltke, the formation of its ethos and its pioneering work in adapting German military thought to the railway age is to be found in A. Bucholz, *Moltke, Schlieffen and Prussian War Planning* (Berg), 1991, pp. 39-57.

8. A useful summary of the early development of mechanized and armoured forces in the German Army is to be found in W. Heinemann, 'The Development of German Armoured Forces', in J. P. Harris and F. H. Toase, *Armoured Warfare* (Batsford) 1990, pp. 51-70.

9. Ibid., pp. 52-4.
10. Ibid., p. 53.
11. R. M. Ogorkiewicz, *Armour* (Atlantic), 1960, p. 209.
12. Guderian, op. cit., pp. 22-4.
13. Ibid., p. 24.
14. Ibid., pp. 20-2. A number of Guderian's previous published writings are referred to in the notes to this edition of *Achtung – Panzer!* His first article in *Militär-Wochenblatt* appears to have been 'Truppen auf Kraftwagen und Fliegerabwehr', No. 12, 1924, pp. 305-6.
15. Guderian, *Panzer Leader*, p. 23-4.
16. Ibid., p. 25.
17. D. Bradley, *Generaloberst Heinz Guderian und des Entstehungsgeschichte des modernen Blitzkrieges* (Biblio-Verlag) Osnabruck), 1978, passim.
18. Macksey, op. cit., pp. 50-2.
19. D. Irving, *The War Path* (Michael Joseph) 1978, pp. 22-3.
20. Harris and Toase, op. cit., p. 57. David Irving, however, places Hitler's visit to Kummersdorf in February 1935. See Irving, op. cit., p. 45.
21. The figures for September 1939 are six panzer divisions with a seventh in the process of creation. At the same time there were four motorized and 86 unmotorized infantry divisions, four light divisions and three mountain divisions. M. Cooper, *The German Army* (Macdonald and Jane's), 1978, pp. 162-3.
22. Guderian, *Panzer Leader*, p. 38.
23. Ibid., pp. 32-3.
24. Ibid., pp. 32-9.
25. Ibid., p. 38.
26. A. Hitler, *My Struggle* (Hurst and Brackett), 1933, pp. 90-2.
27. A recent account of the genesis of the tank which indicates that Swinton's role was by no means central is to be found in R. Ogorkiewicz, *The Technology of Tanks* (Jane's), 1991, pp. 3-7.
28. Some of Churchill's early suggestions in the field of trench-crossing vehicles were completely impracticable. D. Fletcher, *Landships* (HMSO), 1984, p. 4.
29. Cooper, op. cit. pp. 115-17.
30. For a summary of the development of British armour in the inter-war period see Harris and Toase, op. cit., pp. 27-51.
31. J. Mearsheimer, *Liddell Hart and the Weight of History* (Cornell), 1988, pp. 189-91.
32. Guderian, *Panzer Leader*, p. 142.
33. Macksey, op. cit., p. 69 and Guderian, *Panzer Leader*, p. 89.

PREFACE

If the fundamental principles of combat are identical for all arms of service, their application is strongly conditioned by the technical means that are available.

Even now opinions are sharply divided concerning the employment and operations of tanks. This should occasion no surprise, since all armies are burdened by a strong, if not limitless, power of inertia. The lessons of the World War point without exception to the importance of concentrating large masses of tanks on the decisive spot – a practice which also happens to correspond with the principle of forming a principal axis of effort (*Schwerpunktbildung*). To many observers, however, the experiences of the war fail to offer convincing guidelines, and not least because the means available to the defence have shown a quantitative and qualitative improvement in the intervening years.

One thing remains clear, that every technical means of combat – tanks included – must be developed to the farthest limit of its potential. It follows that we should not restrict our opportunities out of a regard for tradition. On the contrary, we must take our lead from the new weapons in question. What we carry from the past must be developed farther, and if necessary changed, by the possibilities which now lie before us.

With these considerations in mind I express the hope that the present book will contribute to the clarification of our thinking.

General der Panzertruppen Lutz

OVERVIEW OF THE WESTERN FRONT

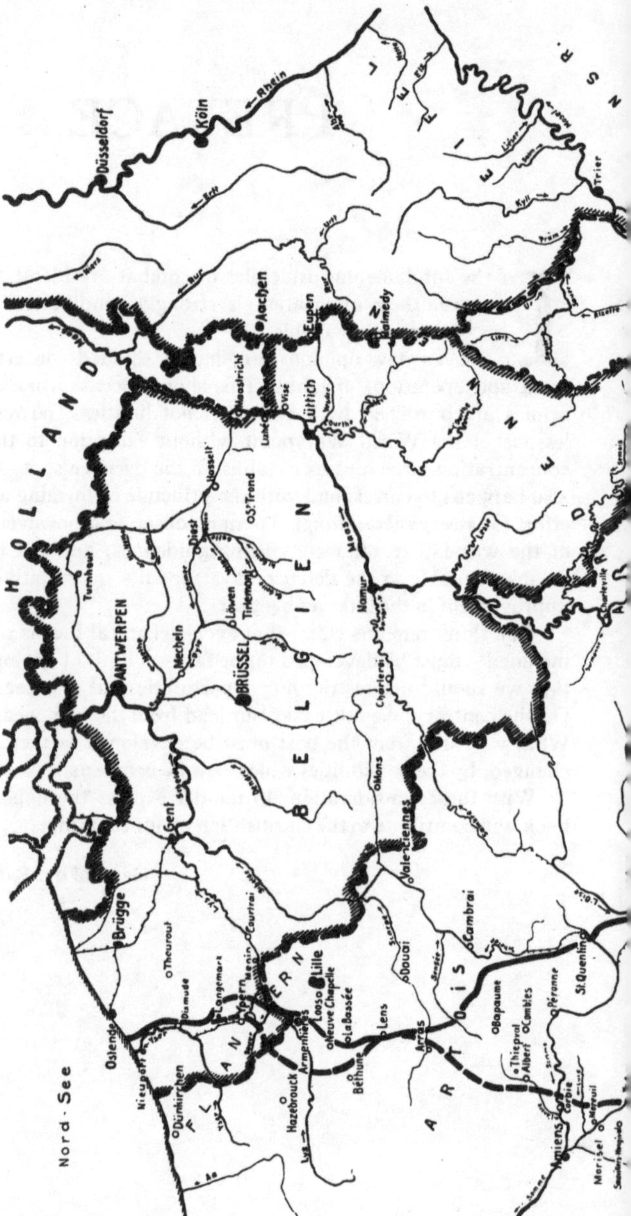

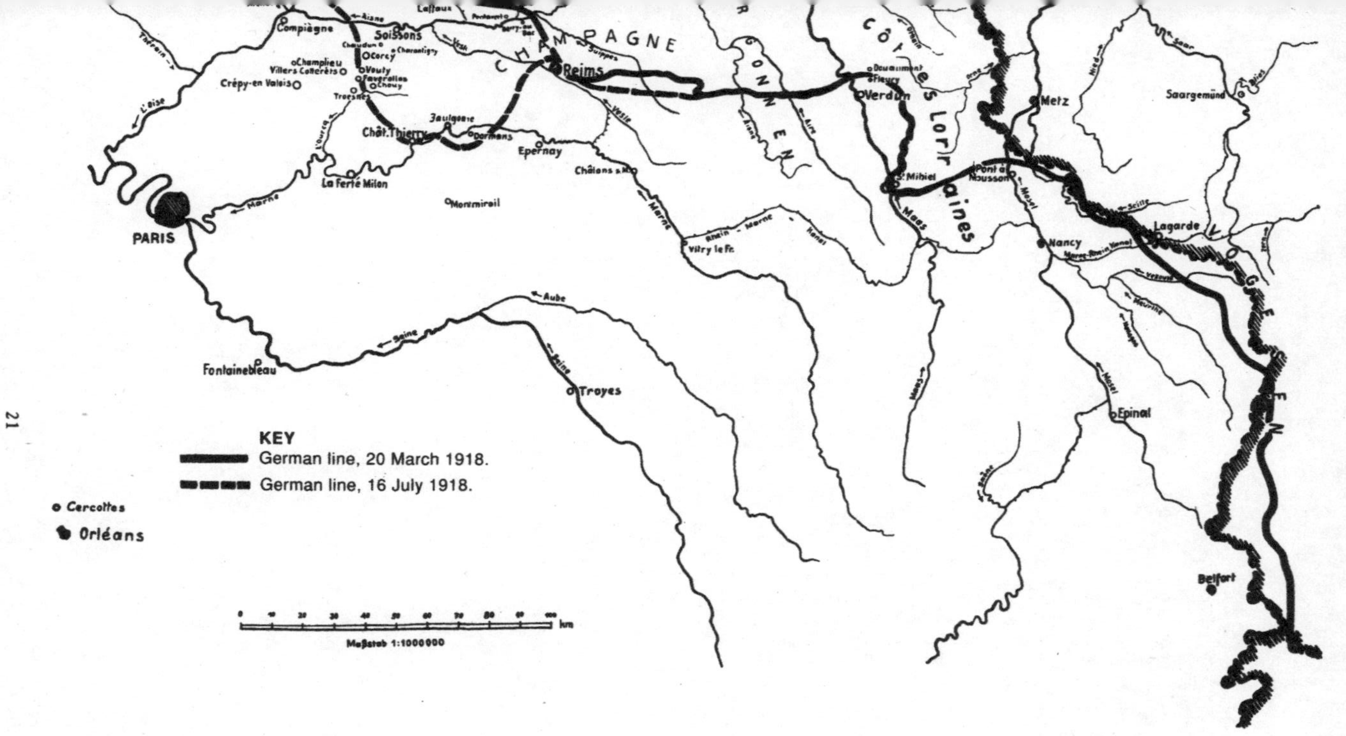

KEY

German line, 20 March 1918.

German line, 16 July 1918.

INTRODUCTION

We live in a world that is ringing with the clangour of weapons. Mankind is arming on all sides, and it will go ill with a state that is unable or unwilling to rely on its own strength. Some nations are fortunate enough to be favoured by nature. Their borders are strong, affording them complete or partial protection against hostile invasion, through chains of mountains or wide expanses of sea. By way of contrast the existence of other nations is inherently insecure. Their living space is small and in all likelihood ringed by borders that are predominantly open, and lie under constant threat from an accumulation of neighbours who combine an unstable temperament with armed superiority. Some powers may have considerable natural resources and colonial territories at their disposal, and derive therefrom a considerable degree of independence both in war and peace; others, who are no less viable and may, indeed, often be larger in terms of population, will possess a very restricted base of raw materials and few if any colonial territories. Because of this they live in a state of continual economic stress and are in no position to sustain a long war.

The pattern of historical development, together with the lack of insight on the part of nations who live in a state of superiority, have created a condition of crisis for those nations who are unable to tolerate a long period of hostilities, with all its attendant economic privations. Such nations have been forced to consider what means may best conduce to bring an armed conflict to a rapid and tolerable end. If we are impelled to throw ourselves into this debate it is because we remember all too vividly the state of famine that was occasioned in the Central Powers by the war, and by the blockade which was so cruelly prolonged beyond the Armistice.[1]

Leaving aside other mistakes on the part of the political and military leadership, we must recognize that in 1914 the offensive power of our army was not sufficient to bring about a rapid peace. That is to say that our armament, equipment and organization did not permit us to pose a material equivalent to the enemy's numerical superiority. We believed that we possessed a moral superiority of our own, and indeed we were probably right. But this superiority was not enough to win the day. The moral and intellectual condition of a nation may certainly prove of decisive importance on its own account, but all due attention must also be paid to material considerations. When a nation has to reckon with a struggle against superior forces on several fronts, it must neglect nothing that may conduce to the betterment of its situation.

All of this may appear self-evident; but military literature is replete with statements that indicate that many people believe that we may embark on a new war with the weapons of 1914, or at best with those available in 1918. Many authorities consider themselves forward-looking when they bring themselves to admit the value of the new weapons which appeared towards the end of the war – as auxiliaries of the old ones. This is a narrowing and negative concept. Fundamentally these men are unable to break free of the memories of positional warfare, which they persist in viewing as the form of combat of the future, and they are incapable of summoning up the necessary act of will to stake everything on a rapid decision. In particular they are blind to the prospects that are opened by a full exploitation of the internal combustion engine. 'It is a love of comfort, not to say sluggishness, that characterizes those who protest against revolutionary innovations that happen to demand fresh efforts in the way of intellect, physical striving and resolution.' Hence we encounter the outright assertion that motorized and mechanized weapons represent nothing revolutionary or new, and dismissive comments on the lines that, their 'single' chance of success came and went in 1918, that they have had their day, and that one may content onself with standing on the defensive. We could cite other statements equally smug and negative. But the facts speak otherwise. 'One thing is certain: the replacement of muscle power by this new machine will lead to one of the mightiest technical – and therefore economic – transformations that the world has ever seen. Far from being at the summit of these developments, I believe we are only at the beginning' (Adolf Hitler at the opening of the Automobile Exhibition, 1937).

Such revolutionary economic changes must lead, as always, to military changes of a corresponding order; it is a question of making sure that military developments keep pace with the technical and economic ones.[2] This is only possible if we welcome the develoments in question whole-heartedly, and not just pay lip service to them. Such a whole-hearted affirmation, which is a precondition of promoting these developments, demands that we should assess the actual effect of weapons in the last war, beginning with the weapons and arms of service which took the field in 1914, and proceeding to those – most of them, unfortunately, wielded by the enemy – with which we had to reckon in 1918. We next have to survey the developments that took place in foreign countries while we ourselves laboured under the restrictions imposed by the Versailles *Diktat* (Treaty of Versailles), and finally we must use our investigations as a means of drawing conclusions for the future.

This book does not set out to present a history of the technical development of tanks; such a work would demand a specialized and comprehensive treatment by expert authorities. I touch on the technical unfolding of this new weapon only as far as seems necessary to explain the course of military events. In this volume I am striving very much more to describe the development of the tank arm from the viewpoint of the soldiers who must wield it; my work is therefore concerned chiefly with combat tactics, and the operational exploitation of tactical success. If the tactical lessons are drawn from the events on the Western Front between 1914 and 1918 it is because that theatre was where the main decision of the war was reached, and where our strongest enemies and we ourselves deployed the most potent and the most modern means of combat. It was here that these weapons made their first appearance in warfare, and it is with them that we must chiefly reckon in the future.

The reliability and comprehensiveness of the sources concerning these weapons leave, unfortunately, a great deal to be desired, and they make the task of impartial assessment all the more difficult. Twenty years have passed since these devices made their début, and it is high time that official historiography got down to describing how they performed. Until that time we will have to make do with unofficial researches that have been conducted under difficult conditions, and which are riddled with lacunae.[3]

My aim in this book is to inspire veterans and young soldiers alike to reflect on these matters, to look into them more deeply, and then proceed to purposeful action; I hope also that the work will convey to our able-bodied youth an image of our panzer forces, and teach them how to master the technical achievements of the present age and put them at the service of the Fatherland.

Notes

1. Guderian's theme in the first two paragraphs of the book is Germany's strategic position and its consequences for her conduct of war. The argument would be familiar to most German readers. Germany lies in the centre of Europe with potential enemies on two fronts. She lacks adequate quantities of certain important raw materials and, being inferior at sea, she is subject to blockade. Therefore she cannot afford long indecisive wars and must take steps to eliminate enemies quickly in short, sharp campaigns. Guderian's originality lies neither in his analysis of the strategic position nor in seeing the desirability of swift, decisive victories, but in his ideas on the means of accomplishing these. In other words his contribution is in operational and tactical not strategic thought.

2. Guderian's statement of the necessity for adapting military doctrine to keep pace with economic and technical change is very much in keeping with the traditions of the German General Staff. Under the elder Moltke (its Chief 1857–87) it had led Europe in the study of the impact of railways on war. See P. Paret (ed.), *Makers of Modern Strategy* (Oxford), pp. 287–88. Guderian is here arguing the necessity of adapting military doctrine to the era of the internal combustion engine.

3. Guderian is right to point out the weaknesses of his source material. Remarkably, considering the enormous outpouring of historical writing since 1937, we still lack documented, scholarly accounts of many aspects of the subject matter with which Guderian is dealing. For example there is no fully documented book on the development and use of tanks by the British in the First World War. We are still largely dependent on the memoirs of participants and on an undocumented regimental history.

1914. How Did Positional Warfare Come About?

1. Lances Against Machine-Guns

The August sun shone down mercilessly on the low rolling country as it stretched from the north-west bank of the Meuse at Liège westwards in the general direction of Brussels. Between 5 and 8 August 2nd and 4th Cavalry Divisions, under General von der Marwitz, crossed the Meuse at Liège on the Dutch-Belgian border, and on 10 August encountered numbers of the enemy who were dug in east and south-west of Tirlemont. The Germans decided to outflank them to the north, and the two divisions were temporarily disengaged and pulled back on 11 August to the area east of Saint-Trond, where they rested. The exertions of these first days of the campaign were extremely demanding, and as early as 6 August the Germans began to run alarmingly short of oats for their horses. The earlier probing actions had established that the Belgian troops had withdrawn from Ligne on Tirlemont, and that the Belgian Army would not deploy for action in front of the line Louvain–Namur. Strong forces and fieldworks were identified behind the line of the Gette running from Diest through Tirlemont to Judoigne.

From Tirlemont downstream the Gette itself formed an obstacle in its own right and was augmented by the wet water meadows and a number of drainage ditches; north of Haelen they emptied into the Diemer, which flowed from the east by way of Hasselt. Downstream from that location the Diemer measured ten metres wide by two metres deep. Visibility was restricted by rows of trees and hedges, and many of the built-up areas and fields were divided by wire fences. North of the Diemer a canal (again ten metres wide by two metres deep) ran almost due north from Hasselt to Turnhout, where the Greater and Lesser Nethe flowed into the mighty, fortified city-port of Antwerp on the Schelde.

Altogether the terrain and the way it had been developed posed considerable difficulties in the path of cavalry when it was advancing along the roads; these difficulties became downright intolerable as soon as the Germans tried to make their way cross-country on horseback.

On 12 August General von der Marwitz sought to outflank the defended sector of the Gette by a move northwards in the direction of Diest. With this intention he set 2nd Cavalry Division in motion by way of Hasselt, and 4th Cavalry Division (reinforced by 9th Jäger Battalion and the cycle company of 7th Jäger Battalion) by way of Alken and Steevort to Haelen, while reconnaissance patrols crossed an imaginary line running from Hechtel to Tirlemont by way of Beeringen and Diest. Ten Cavalry Brigade of 4th Cavalry Division remained at Saint-Trond to protect the left flank, with a reconnaissance squadron posted further south-west at Landen.

Second Cavalry Division seized a number of weapons in Hasselt, and after some delay it marched at about noon to Steevort, on the Haelen road. Fourth Cavalry Division had already arrived at the same location, which meant that both divisions were now formed up in line ahead on a road which lay uncomfortably close to the enemy front. During the march General von der Marwitz ordered 4th Cavalry Division to open the passage of the Gette at Haelen, with 2nd Cavalry Division taking the lead by pushing to Herck-la-Ville and securing the ground northwards in the direction of Lummen. The patrols reported that the crossing at Haelen was held by the enemy,

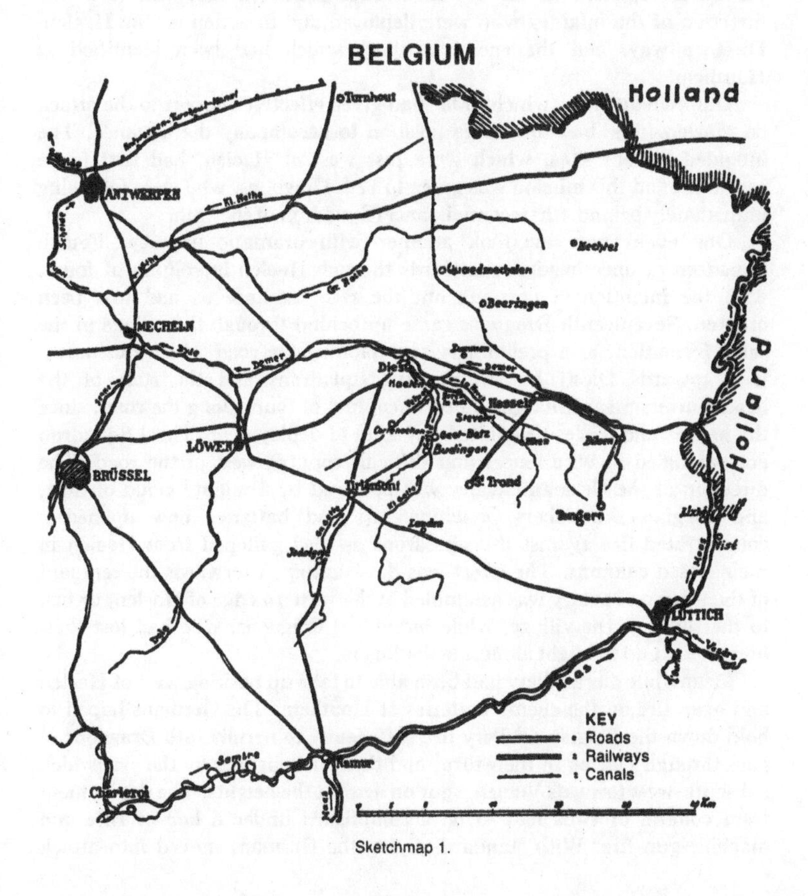

Sketchmap 1.

and General von Garnier accordingly brought his artillery into position west of Herck-la-Ville, while he deployed the reinforced 9th Jäger Battalion on both sides of the Haelen road, and set 3 Cavalry Brigade the task of outflanking the enemy to the south. Towards 1300 the Jägers seized the damaged bridge over the Gette and penetrated to the western end of the village of Haelen. It was now that the enemy artillery opened up – setting buildings ablaze, sweeping the village street from end to end, and inflicting the first German casualties. The Germans now recognized that the heights west of Haelen were occupied by the enemy.

Meanwhile 3 Cavalry Brigade (2nd Cuirassier Regiment and 9th Ulan Regiment) with the help of their pontoon wagons had made a passage of the Gette at Donck, south of Haelen, and were in the process of crossing the river. Seventeen Cavalry Brigade (17th and 18th Dragoon Regiments) had moved up immediately east of Haelen, and had designated 4th Squadron of the latter regiment as the reconnaissance squadron and sent it in the direction of the infantry who were deployed and in action on the Haelen-Diest railway, and the enemy artillery which had been identified at Houthem.

Our own artillery, which so far had given effective support to the attack on Haelen, now had to change position to accompany the advance. The intended battery sites, which were just west of Haelen, had first to be captured, and this mission was given to 17th Dragoons, who were following immediately behind 4th (reconnaissance) Squadron of the 18th.

One event now overtook another with dramatic urgency. Fourth Squadron at once headed westwards through Haelen in column of fours, with the intention of carrying out the reconnaisance as had just been ordered. Seventeenth Dragoons came up behind through the village in the same formation, as a preliminary to exploiting the road which ran north-west towards Diest. Its two leading squadrons and the staff of the headquarters meanwhile remained in column of fours along the road, since the hedges and fences prohibited any kind of deployment. Third Squadron got entangled in wire fences and difficult country west of the road. The direction of the German cavalry was betrayed by a mighty cloud of dust, and Belgian skirmishers, machine-guns and batteries now opened a concentrated fire against the squadrons as they galloped from Haelen in their closed columns. The effect was devastating. Afterwards the remnant of the German cavalry was assembled at the western edge of Haelen, or just to the south of the village, while individual dragoons, who had lost their horses, kept up the fight alongside the Jägers.

Meanwhile our artillery had been able to take up position west of Haelen and open fire on the enemy batteries at Houthem. The Germans hoped to hold down the Belgian artillery fire sufficiently to permit 18th Dragoons to pass through Haelen in their turn, and then, debouching by the exit which led south-west towards Velpen, spur on against the heights. The deployment from column of twos had to be accomplished under a hail of rifle and machine-gun fire. With standards flying the Germans moved into attack formation with two squadrons making up the first line, and the third in echelon to the left rear, and in the process the horsemen rolled over the foremost lines of the enemy skirmishers. Then, however, the attack was shattered by an outburst of violent defensive fire among a zone of hedges and barbed wire fences. The German losses were extremely heavy.

While these events were unfolding 3 Cavalry Brigade met its own fate. The brigade had made a successful passage of the Gette at Donck, and it was there that it received the order to sweep onwards and capture the enemy artillery. Without losing a moment the regiment of Königin Cuirassiers galloped through Velpen with a first line of three squadrons; this charge too was beaten off with severe losses. The regimental commander renewed the attack with the third squadron, which was still intact, and the remnants of the first two squadrons. It was all in vain, and a third and last effort proved to be no more successful.

Just to the right of the cuirassiers 9th Uhlan Regiment was attacking in the direction of Tuillerie-Ferme, with two squadrons in its first line and two in the second; after the first line collapsed the second took up the attack, only to meet the same fate. After the failure of the cavalry assault the push was continued in the direction of Houthem by the Jägers who, from 1400, had the support of skirmishers from the Leibhusaren Brigade, who had dismounted for combat on foot. The Germans took Liebroek to the north, and Velpen to the south.

However the fact remains that for the first time in the war an attempt had been made to charge modern weapons with cold steel, and the effort had miscarried.

What had the enemy being doing?

From 0500 on 10 August the Belgian cavalry division had been positioned behind the Gette between Budingen and Diest, with the purpose of holding that sector and pushing reconnaissance patrols in the direction of Tongres, Beeringen and Quaedmechelen. The villages of Budingen, Geet-Betz and Haelen had been put in a state of defence, and all the bridges over the Gette had been destroyed except for the two at Haelen and Zelck, and these had been prepared for demolition. Enemy cavalry patrols had been beaten off. On the morning of 12 August strong forces of German cavalry were detected on the march for Hasselt. A request for reinforcements was accordingly made to the Belgian high command, whereupon 4 Infantry Brigade was put at the disposal of the cavalry division and had set out in the direction of Cortenaeken at 0815 on the day of the battle; without pausing for rest the leading reinforcements accomplished a forced march of twenty-one kilometres in crushing heat, and at 1600 they arrived on the scene of action in the form of four weak battalions of infantry and a battery of artillery. This battery was the first element of the Belgian forces to arrive, and after planting itself at Loxbergen it had taken up the duel with the German batteries.

The Belgian positions at the beginning of the action are shown on Sketch Map 2. By 1600 most of the reserves had been fed into the infantry fight.

After his 4 Infantry Brigade had arrived, the Belgian divisional commander, General de Witte, resolved on a counter-attack against Haelen on both sides of the Gette. The assault was broken at Velpen by the fire of the German Jägers, machine-guns, the Leibhusaren and the artillery.

Towards 1830 General von der Marwitz broke off the action and assembled his forces east of the Gette.

Four German cavalry regiments had taken part in the attack, and their losses amounted to 24 officers, 468 men and 843 horses; total Belgian losses came to ten officers, 117 men and 100 horses.

What is notable about the action at Haelen? It represents a commitment of cavalry in considerable force (if not simultaneously) against defending infantry and artillery. We see essentially the same outcome in the larger attacks which were launched in the face of enemy fire on the other fronts, such as those of the Bavarian Uhlan Brigade at Lagarde on 11 August 1914, or 13th Dragoon Regiment at Borzymie on 12 November. This indicates that the example of Haelen holds true for many other actions.

The original task of General von der Marwitz had been to advance against the line Antwerp–Brussels–Charleroi so as to pin down the Belgian, British and French forces in Belgium. Nowadays it is fair to pose the question why von der Marwitz, once the Belgians had been identified behind the Gette south of Diest, did not attempt to strike out north of the Diemer. If he had succeeded in pinning down the Belgian north wing, he

ACTION AT HAELEN, 12 AUGUST 1914

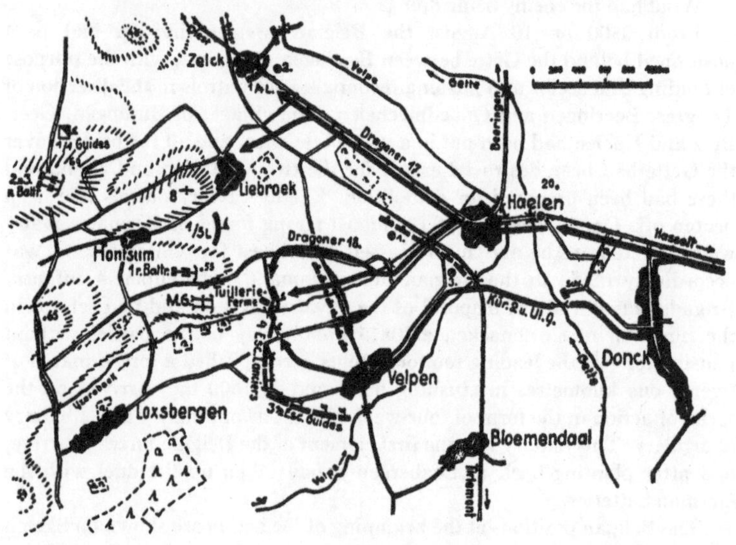

Sketchmap 2.

could have executed the reconnaissance at least as far as the line running from Antwerp to Brussels and operated against the enemy flanks – whether by an envelopment beyond the Diemer in co-operation with the corps of First Army, or by making it difficult for the Belgians to break free in the direction of Antwerp by barring the crossings of the Diemer and the Dyle. It is also reasonable to ask why the attack on Haelen and the Gette, once it had been decided, was not carried out on a broader front by the whole of the cavalry corps simultaneously, and initially at least by a dismounted assault, so as to win a sufficiently wide bridgehead, break the cohesion of the defence, and then exploit the speed of the horses to pursue the shattered enemy.[1]

We discover the answer to these questions when we identify the notions by which the cavalry in Germany – and indeed in foreign countries as well – were educated, equipped and trained.

These ideas are expressed most clearly in the last set of pre-war regulations. They are dated 1909, and the section on tactics opens with the words: 'Mounted action is the predominant way in which cavalry fights.' Ignoring the lessons of one-and-a-half centuries of warfare, the authors of the regulations adhered not only to the spirit, but to a considerable extent also to the form, of the battle tactics of von Seydlitz [Frederick the Great's cavalry commander], and they believed that they could brush aside all the intervening developments which had been dictated by the accelerating march of technology. The equipment and weapons reveal a hankering after the great cavalry battles of the past, while the training put an excessive emphasis on riding school perfection, drilling in close formations and the mounted attack.

We have seen the implications for the commanders and troops in the first actions of the war. We have noted the price that had to be paid in blood. In all probability the reports that Belgian cavalry were making a stand at Haelen led the Germans to believe that the enemy were indeed drawing themselves up for mounted combat; the reports also inclined the Germans to underestimate the endurance and tactical effectiveness of the Belgian cavalry in dismounted action. Here, as elsewhere, the result was a bloody repulse which sapped the trust of the troops in their leadership, while exaggerating their respect for the power of the enemy.

Von Schlieffen painted a picture of the modern battlefield as early as 1909, and it is as valid now as it was then. 'Not a horseman will be seen. The cavalry will have to accomplish its tasks out of range of the infantry and artillery. Breech-loaders and machine-guns will have banished the cavalryman quite mercilessly from the battlefield.'

On the question of operational reconnaissance by cavalry, the following verdict is delivered by the Reichsarchiv's Official History: 'It became only too clear at the outset of the war, and along the whole battlefield, that in peacetime altogether excessive hopes had been pinned on strategic reconnaissance by large bodies of cavalry. As a general rule the probing cavalry managed to identify the enemy outpost line, but they were never able to

FIRST BATTLE OF YPRES, 20 OCTOBER 1914

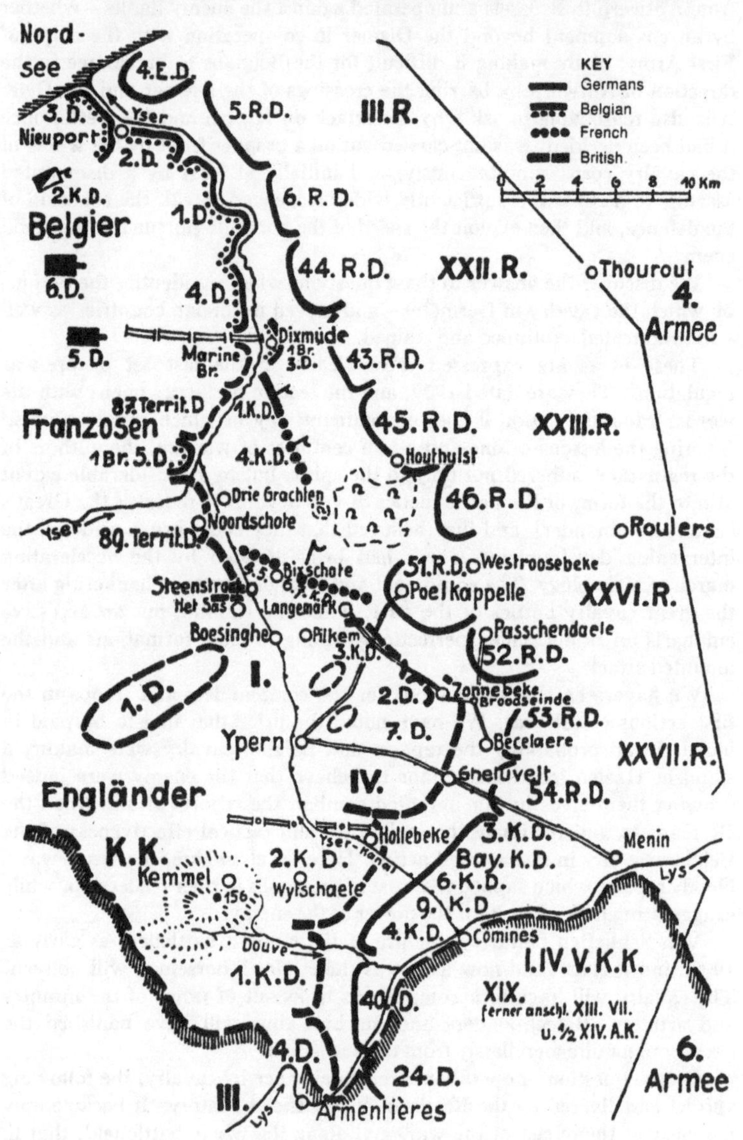

break through and ascertain what was going on in the enemy rear areas.' (Reichsarchiv, I, 126).[2] In 1914 the high command overestimated the effectiveness of operational reconnaissance by cavalry, but they neglected the new dimension of reconnaissance offered by aircraft, even though some machines already had a range of more than four hundred kilometres. The new aviation forces were therefore consigned to the headquarters of the individual armies and corps, and the high command therefore received only a patchy vision of the enemy deployment (Reichsarchiv, I,127).

THE MARCH OF THE INFANTRY TO THE SACRIFICAL ALTAR

The finest army in the world had flooded like a wall of water across the Meuse and deep into the enemy country to the south. Two months later, when the leaves were falling in the autumn of 1914, the grey tide was ebbing – mistakes on the part of the high command, together with heavy losses and logistic problems had combined to bring about an equilibrium of forces along the whole long front between the northern French border at Lille and the Swiss mountains. Against this background the German command resolved to deliver a powerful new blow by fresh forces. It was to be struck in October by our far right wing in Flanders, and the aims were to prevent the front from sticking fast, and to seize the chance of victory before it could elude us.

Hundreds of thousands of volunteers had flocked to the colours at the time of the mobilization – lads brimming with enthusiasm, and older men who were ready for any sacrifice that might be demanded of them. Now, after a sketchy training which rarely lasted more than six weeks, they were hastening to the newly formed corps and divisions on the various fronts. The new Fourth Army was composed of XXII, XXIII, XXVI and XXVII Reserve Corps, together with III Reserve Corps and 4th Ersatz Division from Antwerp (which already had some combat experience) and a reasonably strong complement of heavy artillery by the standards of the time. On 17 October the army began to move from its start-line running from Bruges to east of Courtrai, and made for the line of the Yser between Nieuport and Ypres. In all probability no German soldiers had ever gone into battle with such enthusiasm and élan as the men of these young regiments.

On 19 October contact was made with the enemy along the whole frontage of the army, and on the next day the combat in Flanders, the First Battle of Ypres, began to unfold. In addition to Fourth Army, attacking on its sector north of the Menin–Ypres road, the offensive was opened simultaneously by the formations adjoining to the south, namely the battle-tested right-flanking corps of Sixth Army (V, IV and I Cavalry Corps; XIX, XIII, VII and half of XIV Corps, with II Cavalry Corps behind) which had the task of striking west and facilitating the advance of the new corps of Fourth Army over to their right.

We now turn to some features of the area designated for the attack. We first trace the course of the Yser from its mouth on the coast at Nieuport upstream by way of Dixmude to Noordschoote, and so to the Yser Canal, which runs by way of Steenstraate, Boesinghe, Ypres and Hollebeke to Comines. On each side of the Yser, from Dixmude to the sea, extends a deep polder landscape, some of it lying below sea level, which is intersected by numerous ditches and canals. The level of the water, and if necessary the inundation by the sea, is regulated by a system of sluices which has its most important concentration at Nieuport. Mount Kemmel, 156 metres high, stands south of Ypres. From here a circlet of ridges extends in a flattish arc through Wytschaete, Hollebeke, Gheluvelt, Zonnebeke and Westrobeeke in the direction of Dixmude. This feature was particularly important for artillery observation in this otherwise flat landscape, where visibility was heavily restricted by the numerous farmhouses, hedges, copses and villages. In such terrain the direction of combat, expecially where inexperienced troops were concerned, was severely degraded.

On 20 October the troops of the new regiments assaulted Dixmude, Houthulst, Poelkappelle, Passchendaele and Becelaere, with the words of 'Deutschland über Alles' on their lips. Their losses were extremely heavy, but the gains were reasonable enough.

On the night of 20/21 October came the order to continue the attack over the Yser. The village of Langemarck and the crossroads at Broodseinde lay in the path of the troops. The German artillery fired to supposedly destructive effect, after which the young regiments renewed the attack. Reserves pushed forward, filling the gaps in the forward lines as they were thinned out, but the result was only to increase the losses. The officers took a personal part in fighting, but this did nothing to diminish the carnage caused by the enemy fire; the losses became uncountable, and the offensive potential of the Germans drained away. A new bid was made to capture Langemarck on the 22nd. Not only did it fail, but the Germans came under counter-attacks which showed that the resolution of the enemy was still unbroken. Meanwhile a push farther to the north-west got as far as the eastern edge of Bixschote, while an assault yet farther north reached the gates of Dixmude. The fighting on 23 October rewarded our frightful losses with no gains whatsoever, and our troops had to call for entrenching tools and dig in. 'By the evening of 23 October, after four days of battle, the first onslaught on the Yser Canal by the new corps had been brought to a standstill.' (Reichsarchiv, V,317.)

The enemy had not been particularly strong, and yet our infantry did not have the offensive strength to overcome them, even with reasonably powerful support from our heavy artillery. The most noble self-sacrifice, the glowing enthusiasm and energetic command – all proved unavailing. Nowadays it is fashionable to allege that it had been a mistake to choose this vital sector of the front as the place to commit the young and inexperienced reserve corps, with their ageing leadership and patchy training. People who argue along these lines fail to grasp the fact that the

First Battle of Ypres shows that infantry lacks the striking power to overcome an enemy, even when that enemy happens to be numerically inferior. I am willing to concede that troops with combat experience might well have gained the same results with fewer casualties; but it is doubtful whether their losses would have been significantly lower, and more questionable still whether they would actually have scored a victory. We have to bear in mind that the inexperienced troops were not the only ones to be committed to that last great offensive of 1914 in those streaming wet days of October. The crack III Corps was in action on their right, and the combat-tested divisions of Sixth Army were fighting on their left; their opponents were not notably stronger or more battleworthy than the troops that faced the newly trained divisions, and yet the Germans achieved no more on these sectors than elsewhere. The situation of the two sides is shown on Sketch Map 3.

The Germans ran through most of their small stock of artillery ammunition, and from 24 October the attack petered out in individual engagements, and finally almost literally drowned in floods of water. Two final attempts were made to exorcise the spectre of deadlock on the Western Front. They concerned brigades and divisions of picked troops which were withdrawn temporarily from the front line for this purpose, and yet both bids failed in a series of bloody combats.

From 30 October to 3 November XV, II Bavarian and half of XIII corps were formed into an 'Assault Group Fabeck', and five of the divisions were pushed into the attack on a frontage of ten kilometres. The result was yet another bitter disappointment, and it was not ameliorated by the piecemeal commitment of 6th Bavarian Reserve Division, 3rd Pomeranian Division and elements of cavalry.

Finally, from 10 to 18 November, yet further resources of battle-tested troops assaulted the Ypres Salient. Ninth Reserve Division went into action in III Reserve Corps' sector of Fourth Army's front; 4th Jäger Division and the composite Winkler Guard Division were marched to join Sixth Army, where they were incorporated as the 'Assault Group Linsingen' together with XV Corps , which was already in action on the Menin–Ypres road.

On 10 November a number of fresh regiments captured Dixmude and made some progress at Drie Grachten and Het Sas; farther east, however, III Reserve Corps, its 9th Reserve Division in particular, met with no success. The division had been pushed into battle too soon and was badly mauled in the process. On 11 November the Guards and 4th Division attacked on either side of the Menin–Ypres road and made only modest gains; here again the casualties were heavy. The Germans saw that there was no hope of significant progress for the time being, and the commanders of the two armies demanded to be reinforced as a precondition for any further offensives.

The high command consequently instructed Seventh Army to give up one of its infantry divisions to Sixth Army. In the same way Third Army yielded a further infantry division to Fourth Army, accompanied by an infantry brigade from the Army Detachment of General Strantz. The two divisions of reinforcements were represented by their infantry components only and the artillery was left behind. In fact we had plenty of heavy guns, but the high command had to economize on ammunition. In other words the intended attack was deprived of its striking power from the outset – and in the circumstances of the time ammunition should have been given higher priority than numbers of infantry. Fourth Army recognized this fact and simply gave up the idea of any further attacks; Sixth Army threw the Linsingen Group into an assault, which was bloodily repulsed, and then it too made the painful decision to go over to positional warfare. 'On 18 November, between the sea and the Douvre, 27+ German infantry divisions and one cavalry division were facing twenty-two enemy infantry divisions and ten cavalry divisions.' (Reichsarchiv, VI,25.)

Between 10 and 18 November German losses on the sector of the offensive came to some 23,000 men. Over the entire span, from the middle of October until the beginning of December, the losses of Fourth Army amounted to 39,000 killed and wounded and 13,000 missing; a combined total of 80,000 men for the two armies. All in all the First Battle of Ypres cost the Germans more than 100,000 troops, including the flower of its young men and a large part of its resources of leadership.[3] Enemy losses were: French: 41,300, including 9,230 missing (*Les Armées Françaises dans la Grande Guerre*, IV,554); British: 54,000, including 17,000 missing; Belgian:15,000.

In total the losses from August to November 1914 amounted to: Germans, 677,440; French, 854,000; British, 84,575.

3. TRENCH WARFARE AND BARBED WIRE

From the middle of November 1914 all movement ceased along the length of the Western Front. This state of inertia had first set in along the Vosges sector, and it now extended to the coast. And it was in the coastal region that such offensive strength as remained to both parties from the earlier battles had been finally expended in October and November.

What had taken place? On the German side the concentration of offensive power had taken the form of a stream of infantry reinforcements – the young reserve corps of Fourth Army, and various corps, divisions and brigades of infantry from other sectors of the front. They had had an adequate number of artillery pieces, but even at the beginning of the battle they had had at their disposal only limited reserves of ammunition. The whole emphasis of the German attacks was therefore put on the shock effect of the infantryman with his bayonet. The enemy, on their side, were not able to match the Germans in terms of numbers of combatants, and the French, British and Belgians were soon forced on to the defensive. In this form of combat, however, the machine-guns and artillery proved able to withstand the onslaught of superior masses of troops; the hail of fire from modern weapons had shattered the charges of the German lancers in

August, and it now proceeded to do the same to the bayonet attacks in October and November. It was fortunate that the enemy too began to run short of ammunition, otherwise the disproportion in numbers which was experienced in the middle of November, when a quantity of German corps had to be given up for the Eastern Front, would have been still more to the disadvantage of the Germans than it actually was.

The losses which both sides suffered in the flower of their infantry, the lack of ammunition, and the fact that the front line came to rest on its two extremities on the sea and the Swiss mountains – all of these things together forced the armies to resort to the spade and the erection of obstacles. Both sides lived in the hope that the incipient positional warfare would prove to be no more than a temporary condition. Neither side was able to steel itself to the pulling back to terrain better suited to a long-term defensive – they feared that yielding ground which had cost so much blood would be interpreted as an avowal of defeat.

The front therefore congealed along the line of the recent fighting, a state of affairs which necessitated elaborate works of engineering and strong garrisons, and precipitated long-drawn-out combats for patches of ground that were of only local significance. The next step was for the belligerents to develop defensive systems consisting of stoutly built and continuous trenches for the front-line troops and their reserves, supplemented by communications trenches for the movements of reliefs and supplies. These positions were protected by barbed wire obstacles which increased constantly in density and depth. For the time being little use was made of rearward positions. The artillery was located close enough to be able to take both the enemy infantry and artillery under fire – in other words fairly far forward, and in no great depth; to begin with no special measures were taken to provide protection for the guns.

The rival armies then addressed themselves to the business of improving their weapons and equipment. In particular there was a considerable increase in the number of machine-guns – an augmentation which continued until the end of the war, and which raised the status of the machine-gun; from having been an auxiliary weapon of the infantry, it became its chief weapon, and in due course the machine-gun became the prime weapon of the air forces as well. The quantity of artillery was also increased, and the pieces were furnished with a quite unprecedented quantity of ammunition; every possible tube was pressed into service, including a number of elderly models. The work of the engineers also gained in significance; Minenwerfer and hand-grenades came into use, and bunkers, demolitions, inundations and obstacles of all kinds gave the positions more and more the character of fortresses.

Time showed that the Germans suffered more than the enemy from the way they held on to positions that had been dictated by the needs of the moment, regardless of whether they were suited for defence over a period of time. Then again, the habit of packing forces into the front line was attended with a variety of evils. It reduced the number of available reserves, it interfered with their training, it cut short their rest periods, and worst of all it reduced the offensive forces which were desperately needed in the other theatres of war, where they might have settled things more quickly. The situation of our enemies improved with an uncomfortable speed, from our point of view, after the Allies decided to discontinue the sending of large forces to subordinate theatres (as they had done at Gallipoli) and instead concentrated all available reserves of troops, equipment and firepower in France. This is not to say, however, that the decision was necessarily the best. Both sides persuaded themselves that only through the commitment of extraordinary resources could they achieve success in battle on the Western Front, and they each strove by different means to attain this end.

Notes

1. The operations of the major cavalry formations in 1914 was a subject which had fascinated Guderian for years before he wrote *Achtung – Panzer!* These operations had formed an important topic of his military history teaching in the twenties and he had written about them for the military press. See 'Bewegliche Truppenkörper (Eine kriegsgeschichtliche Studie)' in the *Militär-Wochenblatt*, No. 19, 1927, pp. 687–94.

2. This quotation from the German official history about the ineffectiveness of cavalry for strategic reconnaissance, when taken together with von Schlieffen's statement quoted earlier about its lack of combat power, is pretty damning. Guderian was right that as far as western Europe was concerned horsed cavalry was approaching the end of its useful life. But we must make some allowance for Guderian's polemical purpose in writing this book. For a more sympathetic treatment of cavalry in the First World War see S. D. Badsey, The British Army and the Arme Blanche Controversy 1871 – 1921, unpublished PhD thesis, University of Cambridge, 1981.

3. In Great Britain First Ypres is remembered mainly as the battle which bled white the regular British Expeditionary Force of 1914. Guderian reminds us that German losses were also grievous and that the battle was a major strategic defeat for Germany. One of the standard accounts in English is A. Farrar-Hockley, *The Death of an Army* (Barker), 1967.

WAGING WAR WITH INADEQUATE WEAPONRY

1. THE ARTILLERY COMBAT

While in November 1914 the Germans shifted the emphasis on the offensive to the Eastern Front – unfortunately already too late to achieve decisive success – the French high command decided to open an attack in the winter of 1914/15, so as to hinder the Germans from sending further forces to the Eastern Front, and simultaneously exploit the temporary weakness of the enemy in the west. In the words of General Joffre's army field order of 17 December 1914, a decisive battle was to be fought to 'free the land once and for all from the foreign invaders'. The possibility of breaking the vulnerable German lines of communication led to the choice of Champagne as the location for the offensive; additional advantages of this sector were the good communications on the French side, and the uncomplicated nature of the terrain from the viewpoint of the attacker.

After four weeks of preparation the three corps of the French Fourth Army opened the offensive on 20 December. Behind the three leading corps a further corps (I) was held in reserve. The French put their numerical superiority on this sector of the attack at 100,000 men. They had at their disposal nineteen aircraft, 780 artillery pieces of all calibres (which were powerful by the standards of the time) and the usual restrictions on the consumption of ammunition were lifted. Altogether the artillery was going to play a considerably greater role in the preparation and execution of the attack than in former battles.

By committing reserves on this impressive scale the French hoped to break through on both sides of the Suippes–Attigny road. When the actual fighting began, however, the French proved incapable of bringing the infantry of the three attacking corps into action simultaneously. The battle then broke up into individual actions on the part of the various corps and divisions. These combats were prolonged until New Year's Day, because the artillery was frequently unable to destroy the obstructions in front of the German trenches or silence the German machine-guns. Infantry assaults, which were delivered after heavy artillery fire, alternated with days on end of purely artillery action; the trench and mine warfare was incessant, and it was soon necessary to reinforce the engineers. In addition the Germans used every pause in the offensive to launch powerful counter-attacks with the purpose of winning back the sectors of trench they had lost.

At the turn of the year a new corps (III) was placed in readiness as army reserve behind the French Fourth Army's front, and the former army reserve (I Corps) was now inserted on the frontage of the offensive. However the bad weather and strong German counter-blows delayed the execution of the new design. The offensive of the French Fourth Army seemed to have 'broken down into a series of smaller actions, which were devoid of any recognizable coherence, and which were interspersed with pauses which sapped their momentum'. (*Les Armeés Françaises dans la Grande Guerre*, II, 225.) The army commander had recourse to his artillery 'to persuade the enemy that the offensive was still going on'. Cavalrymen were sent to the trenches to free the infantry for offensive action, and the artillery of IV Corps was committed to the battle. The Germans dealt a counterblow on 7 January, which was followed by powerful new French attacks on the 8th and 9th. A further failure on 13 January finally persuaded the commander of the French Fourth Army, General de Langle, to terminate the offensive.

The French gained little, and nowadays they like to put the blame on the weather (*Les Armeés Françaises dans la Grande Guerre*, II, 231ff). Here we have to point out that in those gloomy winter days the weather was just as bad for both sides. It is probably more relevant that, despite their consistent and heavy superiority in infantry, the French were unable to smash enough of the German obstacles, suppress the German machine-gun fire or cripple the German artillery.

Now that the artillery had failed to achieve these tasks, the attacks by the French infantry proved equally ineffective, again despite a considerable numerical superiority. The failure was all the greater since the French neglected to launch simultaneous attacks under unified direction along the whole frontage of the army, but instead favoured local assaults on individually selected points of the German positions. Although the French commander began to doubt the validity of this form of attack, he could think of nothing better than simply committing more *matériel*. Joffre himself emphasized the necessity of a longer artillery preparation and the employment of greater forces on a broader front. He ordered the offensive to be resumed, the artillery battle to continue, and the building of a second defensive system as a safeguard against a possible enemy breakthrough.

The debate on how to launch an attack from a state of positional warfare finally led the French high command, in January 1915, to advocate the massive commitment of infantry in deep formation on a comparatively narrow front, prepared and protected by an overwhelming artillery barrage. General de Langle understood the 'massive commitment of infantry' to mean the deployment for each of the main pushes of 'at least one battalion per division, supported by subsidiary attacks on its flank, so as to pin down the enemy on their entire frontage' (*Les Arméees Françaises*, II, 235).

The preparations for the new offensive encompassed the period from 15 January to 15 February 1915, and its execution lasted, with intermissions, from 16 February to 16 March. It began with two corps in the first line, one of the corps being reinforced by a further division, and the other by a brigade of infantry. The French ranged 155,000 infantry, 8,000 cavalry and 819 guns (including 110 of heavy calibre) against a German force which the French reckon to have numbered 81,000 infantry, 3,700 cavalry and 470 guns (including 86 heavy).

In spite of the two-fold superiority on the side of the attackers, the results on the first days were decidedly modest. As early as 17 February French IV Corps had to be made available from reserve. On the 18th the newly reconstituted French were hit by a German counter-attack which recovered most of what had been lost on the previous days. On 22 February, after further and largely unproductive fighting, General Joffre wrote to the commander-in-chief of Fourth Army: 'It would be unfortunate if your offensive created the impression that we are incapable of breaking through the enemy lines, no matter how powerful the means we employ, and at a time when the enemy strength on the Western Front has been reduced to a minimum.' (*Les Armées Françaises*, II,440.) He accompanied these words with the order to continue the offensive in an energetic way. The attacks resumed on 23 February after a few infantry reinforcements had arrived. The gains were meagre.

From 25 February four French corps stood in the front line, with one (XVI) in reserve. On 27 February this too was sent forward as part of the process of forming a special 'Assault Group Grossetti'. While Joffre held back the rest of this corps in closer reserve, he fed one of its brigades into the offensive on 7 March, supported by eleven detachments of field artillery and fifty heavy guns. Once more the gains amounted to very little.

Joffre now decided to make a last bid to smash through the German front by engaging the main force of XVI Corps. The distinguishing feature

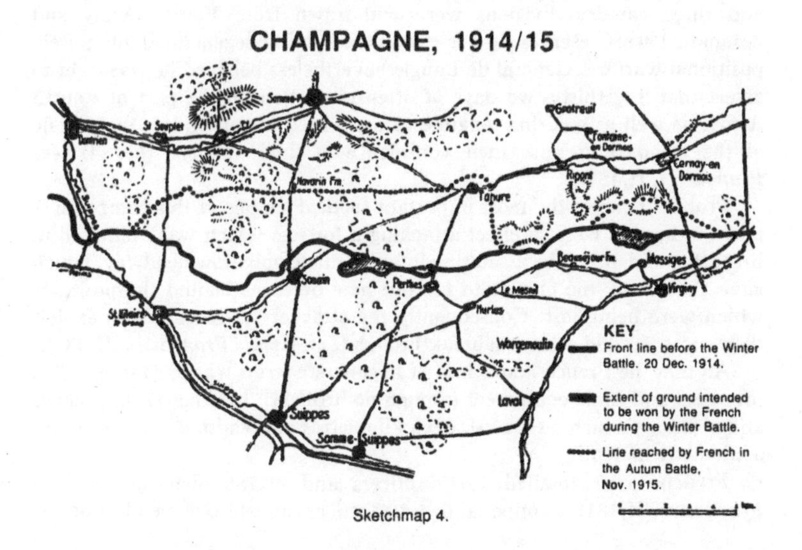

Sketchmap 4.

of this offensive was the deployment of infantry in considerable depth, which signalled from the outset of the battle that the French intended to select a narrow sector for their break-in; the leading assault units were to be relieved after a short period in action, while the attack was to be sustained over a period of days by troops from the rearward elements. Such a deployment renounced any kind of breadth, and helped the Germans to concentrate their defensive resources on the narrow break-in sector.

The last episode of the Battle of Champagne was played out in the fighting which raged from 12 to 16 March. The newly committed troops of XVI Corps achieved no more than those who had already been fighting for a number of weeks. The reserves were expended piecemeal. The commanding general of XVI Corps reported quite justifiably on 14 March that 'in spite of the sacrifices we have made, the offensive will yield unsatisfactory results as long as the assaulting units remain unprotected against close-range flanking fire by the enemy' (*Les Armées Françaises*, II,468). In other words the offensive lacked breadth, the commitment of forces did not match the objectives, and the resources at the disposal of the attackers were not equal to the means available to the defenders. General Grossetti, who had the reputation of being something of a fire-eater, suggested that the French should counter by launching three simultaneous but separate attacks against the objectives which had already been set for him, and only afterwards use the ground thus gained as a base for a larger, more coherent push to the north. The army command approved the proposal. It was supposed to be put into effect on 15 March, but the Germans got in a counter-attack first. On 16 and 17 March the French attacks again gained no more than insignificant local successes. The army commander requested, and obtained, permission to terminate the offensive. Altogether 4½ corps and three cavalry divisions were withdrawn from Fourth Army and designated army reserves. In a few days the fighting degenerated into purely positional warfare. General de Langle nevertheless believed he was right to assert that 'the thirty-two days of offensive action on the part of Fourth Army, as well as securing tangible gains, served to consolidate the morale of the troops and raise their confidence in final victory'. (*Les Armées Françaises*, II,481.)

Probably one of the most important tactical lessons of this campaign is that the French were in effect attacking a fortress which was unlimited in breadth and depth. The assaulting infantry made only slow headway, which gave the enemy the chance to lay out new defences behind the positions which were being lost. Consequently the attackers were unable to exploit their successes and achieve a breakthrough (*Les Armées Françaises*, II,481).

All that the French had achieved in concrete terms was to capture 2,000 prisoners and some equipment (though no artillery), together with trenches and positions which measured seven kilometres in breadth and at most half a kilometre in depth.

French losses totalled 1,646 officers and 91,786 men (*Les Armées Françaises*, II,481), as opposed to 1,100 officers and 45,000 men lost on the

German side. The Germans had taken about 2,700 prisoners. The German positions had been furnished with only a few dugouts and they lacked tactical depth, but they had been held substantially intact against a more than two-fold superiority, thanks to the courage of the troops, the effectiveness of the machine-guns and artillery, and the inexhaustible activity of the engineers. All of this was in despite of the unprecedented outlay of enemy artillery and ammunition – the thunderous 'drum fire' which was to reverberate throughout every battle from now until the end of the war.

Both sides claim victory in the winter battle in Champagne, the first 'artillery battle' of the war. Closer investigation shows that the French had to pay an excessive price for insignificant gains in terrain. The Germans had scanty reserves and feeble artillery with which to hold their ground, which happened to be vital for the stability of their entire Western Front. They nevertheless fulfilled this responsibility in an outstanding way, and we must pay due tribute to Third Army.

The battle demonstrated that the French, for all their undoubted courage and double superiorities in numbers and ammunition, were unable to break through positions which the Germans certainly defended with great obstinacy, but which were not particularly strong in themselves. The reason, once again, was that the defenders always had the time to seal off the locations of the break-ins before the attacking troops, advancing step by step, could exploit their initial successes.

The generals now had to ask themselves, how could they attack with any reasonable chance of success in the future? The obvious thing was to augment the resources already available to the offensive: extend the frontage of the assault so as to pin down more of the defenders at a time, and eliminate local activity on the flanks; one could also build up the quantity of guns and ammunition in the hope of annihilating the defences and obstacles and crippling the enemy artillery.

But novel weapons opened the prospect of something altogether more effective – and poison gas, aircraft and armoured vehicles were already within the grasp of the technology of the time. The Western Front, the most important theatre of the war, appeared to be condemned to deadlock, yet it might prove possible to bring off a major success even here, if some way could be found of employing the new devices – by themselves, or at least in association with one another and with the older weapons – *en masse* and with the advantage of surprise.

Surprise in any event seemed a worthwhile objective, because it might enable one to anticipate countermeasures, secure concentrations *en masse* and enable one's mobile forces to follow up any successes. These desiderata were easy to outline on paper, but less easy to put into effect among the realities of the battlefield. As things turned out, the real or supposed needs of the moment often led to the forces being thrown into action too soon; sheer impatience sometimes led to miscalculations of this kind, and

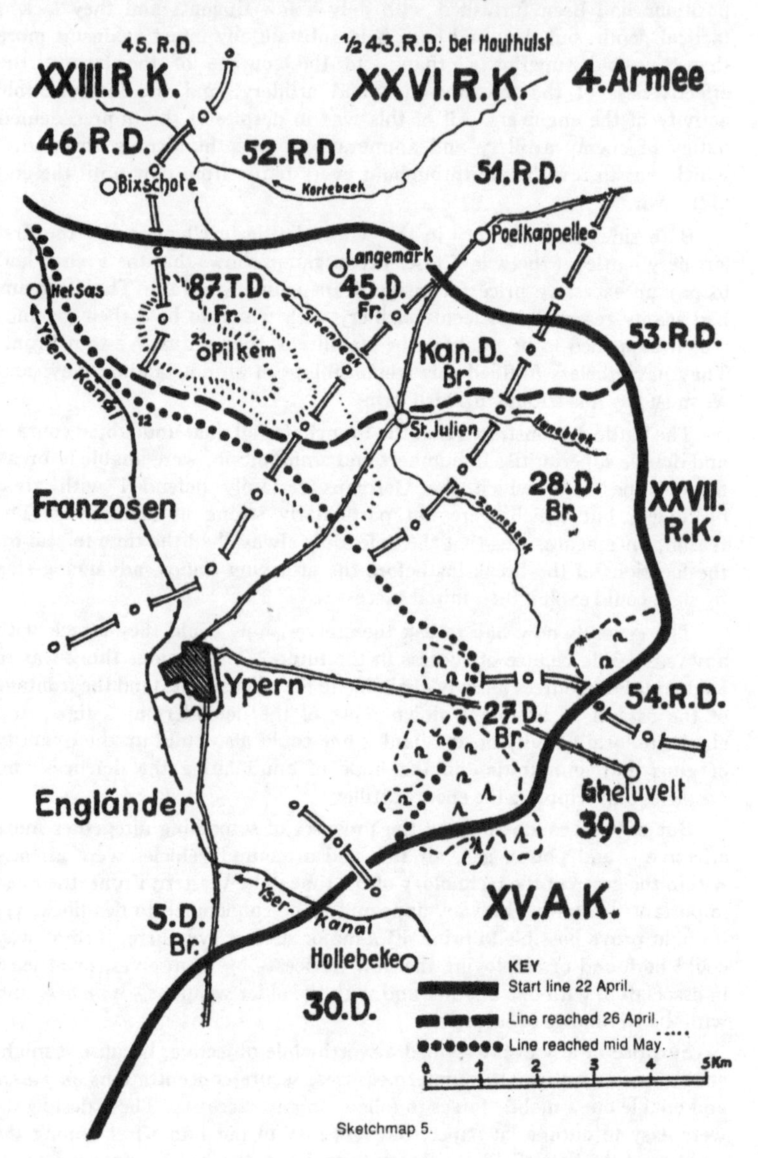
GAS ATTACK AT YPRES, APRIL 1915
45.R.D.
1/2 43.R.D. bei Houthulst
XXIII.R.K.
XXVI.R.K.
4.Armee
46.R.D.
52.R.D.
51.R.D.
Bixschote
Kortebeek
Poelkappelle
Langemark
Het Sas
87.T.D. Fr.
45.D. Fr.
Steenbeek
Yser-Kanal
Pilkem
Kan.D. Br.
53.R.D.
St. Julien
Hannebeek
Zonnebeek
28.D. Br.
XXVII. R.K.
Franzosen
Ypern
54.R.D.
27.D. Br.
Gheluvelt
Engländer
39.D.
Yser-Kanal
XV.A.K.
5.D. Br.
Hollebeke
30.D.
KEY
Start line 22 April.
Line reached 26 April.
Line reached mid May.
0 1 2 3 4 5Km

Sketchmap 5.

sometimes also a feeling of distrust towards such new and unproven weapons.

Since surprise can have a dramatic impact in warfare, it will be rewarding to investigate how the new weapons in question were actually employed, and what kind of impression they made on the enemy. Our inquiry will also show whether the belligerents did any better by employing the conventional alternative – a quantitive increase in the older generation of weapons.

2. GAS WARFARE

We return to Flanders. It was there that a new weapon – poison gas in the form of chlorine – had first come into use in February 1915, when the French employed rifle-gas grenades against the Germans.[1]

'In favourable weather the gas should be released from cylinders in the forward trenches, so as to compel the enemy to abandon their positions.' (Reichsarchiv, VII.53). So the instructions ran. In fact the officers at every level of command, as well as the troops, regarded gas with 'mistrust if not outright rejection' (Reichsarchiv, VII,30). For this reason the Germans at first essayed only a small practical experiment, on the sector of their Fourth Army. The army chose as its immediate objective the heights of Pilckem and the ground to the east, hoping that if all went well the enemy would have to evacuate the Ypres Salient and the Germans would gain the Yser Canal.

At that period chemical weapons came in two forms – gas shells and gas cylinders. The design of the gas shells left a great deal to be desired, however, and there were not enough propellant charges to secure an adequate density at the receiving end. For want of anything better the Germans had to resort to releasing gas from cylinders. These were to be planted in batteries in the forward trenches, and opened when the wind and other climatic conditions were favourable. The heavy reliance on wind and weather proved to be a severe drawback, for it was difficult for the Germans to determine the exact moment of the attack, and this was the fundamental reason why they distrusted the new device. They were also liable to casualties if the weather suddenly changed, or the cylinders were damaged by enemy fire. That was why the Germans decided against using gas on a large scale in their great breakthrough battle which was impending in Galicia, but settled instead on the limited test in Flanders.

Six thousand cylinders containing 180,000 kilograms of chlorine gas were deployed on a frontage of six kilometres on the sectors of XVIII and XXVI Reserve Corps. The cylinders had originally been emplaced further to the west, but they had been moved to the new location as a result of meteorological studies of the wind direction. After a number of delays the long-awaited north wind set in on 22 April 1915, but unfortunately not until the afternoon of that day. All the preparations had been made for an attack at first light, and these now had to be changed. Furthermore the infantry now had to follow the cloud of gas in full daylight, which might expose them to heavy casualties, and deny them the time to exploit any successes. At 1800 German engineers opened the valves of the cylinders, and a dense, whitish-yellow cloud, 600 to 900 metres deep, and extending to the height of a man, was carried on the wind at a speed of between two and four metres a second over the trenches of the French 87th and 45th Infantry Divisions. The enemy were overcome by panic, and after firing a few shots they abandoned their trenches, suffering heavily in the process. The French lost 15,000 men, of whom 5,000 were killed and 2,470 were taken prisoner (including 1,800 uninjured). Losses in *matériel* included fifty-one guns (of which four were of heavy calibre) and seventy machine-guns. Of the two hundred prisoners who were suffering from the effects of gas, only twelve men, or 6 per cent, later died.

By the evening German gains measured eleven kilometres in breadth by a maximum depth of rather more than two kilometres. A gap of some 3½ kilometres had opened between the Yser Canal and Saint-Julien. Regrettably the only force available to Fourth Army to exploit this splendid success was half of 43rd Reserve Division, which was standing at Houthulst, but was too widely deployed and too weak to be able to seize the opportunity while it still existed. Over the following days the Germans enlarged the initial success with the help of repeated gas attacks, and the British were finally compelled to evacuate a considerable part of the Ypres Salient. By the time the battle ended on 9 May German territorial gains covered an area some sixteen kilometres wide by a maximum depth of more than five kilometres. The enemy had overcome their initial panic and very rapidly learned to give themselves some protection by improvising masks. The Germans themselves began to take heavy casualties.

Altogether over the thirteen days of the offensive the Germans lost 35,000 men, and the enemy about 78,000.

When we review the Second Battle of Ypres and the winter battle in Champagne, and compare the respective casualties, the ground that was won and the *matériel* captured, it becomes evident that when a novel weapon is employed with the advantage of surprise it can beat even battle-tested and brave troops who are equipped with modern armaments. As a German one can only regret that the lack of confidence displayed in this new instrument of war – however understandable this might have been – and the consequent failure to hold sufficient reserves in readiness, prevented us from exploiting and expanding the victory with the requisite speed. In future it will no longer be possible to attain surprise through the novelty of gas as a weapon in itself, but only through the place and time we choose to employ it, and the density with which it is applied. This does not rule out as yet unexplored opportunities, especially if gas is used in concert with the older and established techniques of the offensive. In all of this we must bear in mind that the enemy will take protective countermeasures. These will be equally necessary for our own troops, since the enemy will undoubtedly resort to gas warfare as well, and we will also have to reckon with the danger of our own gas blowing back on us.

In the last war releasing gas from cylinders proved to be an unpredictable business, which gave the impulse to a constant improvement in shells that were capable of being fired by artillery or special gas projectors. The result was a weapon that was capable of contaminating a selected area of ground, and crippling every living creature on it, without having to score direct hits by shells or splinters. Among other things it offered the possibility of combating enemy artillery in a way that proved beyond both sides in the battle in Champagne.

On the other hand experiments in gas masks produced a genuine measure of defence against gas. The masks were burdensome to carry, but if they were donned in time they afforded the soldier an immediate protection. This in turn led to a search for substances that were able to penetrate the respirators, and irritate the eyes and the respiratory organs in such a way that the mask had to be torn from the face.

The original forms of gas were designed to facilitate the attack, and were comparatively transient. Very soon, howewver, the combatants began to employ persistent agents which contaminated the ground over a particular length of time, and so facilitated the defensive. The principal substance in question was the so-called 'Yellow Cross', also termed mustard gas. It was not long before chemical weapons became an inevitable presence on every battlefield.

The competition between chemical weapons on the one hand, and masks and other countermeasures on the other, was reminiscent of the struggle between artillery and armour. Both contests were conducted with considerable determination and varying fortune. Finally gas bombs were dropped from aircraft, which gave the range of chemical weapons a further boost.[2]

Notes

1. The French police may have had tear-gas for use in domestic riots since 1912 though it does not seem that they actually used it. The French Army on the Western Front had tear-gas, in rifle grenades, from 1914. Again it is not clear whether it was employed. It is certain that the French were planning a large-scale use of tear-gas for the spring of 1915 but they were overtaken by the German use of chlorine at Ypres in April. The tear-gas the French were planning to use (and may already have used on a small scale) was non-toxic but possibly still in violation of the agreement which France had signed at the Hague Peace Conference in 1899. But in accusing the French of using *chlorine* in February 1915 Guderian would appear to be falsely shifting the blame for initiating real poison gas warfare from the Germans to the French. Whether he was merely mistaken, deliberately lying, or had himself swallowed German propaganda on this topic it is impossible to say. Modern scholarship definitely attributes the first use of chlorine to the Germans. Edward Spiers, *Chemical Warfare* (Macmillan), 1986, p. 14. and L. F. Haber, *The Poisonous Cloud* (Oxford), 1986, pp. 1-40.

2. The most detailed modern account of chemical warfare in the First World War is Haber, op.cit., passim.

THE GENESIS OF THE TANK

1. IN BRITAIN

A number of British officers were impressed by the defensive strength of machine-guns and barbed wire, and as early as October 1914 they were inspired to embark on a counter in the form of an armoured vehicle (the data in this section are drawn principally from Major-General Sir Ernest D. Swinton's book *Eyewitness*, London, Hodder and Stoughton, 1932, pp. 80ff). The model was the Holt Caterpillar Tractor. The running gear was an endless track which gave the machine the potential to crush obstacles, cross trenches and convey its armament under bullet-proof protection into the very midst of the enemy, where it could annihilate the otherwise almost invulnerable machine-guns, and enable one's own infantry to pass open ground without incurring intolerable casualties. These pioneers therefore chose a totally different direction from the Germans, with their chemical weapons; while gas could be employed with no great delay, the British concept had first to be put into tangible form and worked out in practice, which inevitably demanded time.

To begin with the all-mighty Secretary of State for War, Lord Kitchener, rejected the notion of this 'machine-gun destroyer'. In 1898 Kitchener had been the victor at Omdurman, the battle on the Upper Nile where the British had defeated the army of the Mahdi – and this also happened to be the first engagement in which they had employed this murderous weapon. It seems that, under pressure of his work in the World War, he had forgotten just how destructive the machine-gun could be. Perhaps it had slipped Kitchener's mind that not long after the battle he himself had expressed misgivings about what would happen if the British, just like their unprotected native enemies, were called upon to attack hostile machine-guns. It so happened that the experiences of the Boer War did not leave the British commanders with any particularly clear impression of the effectiveness of the machine-gun – this had to wait until the World War.

In December 1914 a memorandum [from Captain Maurice Hankey] finally got through to Prime Minister Asquith. It urged, among other things, that the British ought to build armoured machine-gun carriers with caterpillar tracks. The paper came to the notice of the First Lord of the Admiralty, Winston Churchill, who had very recently been concerned with defending a naval air base at Dunkirk with armoured cars. As wheeled vehicles the armoured cars were road-bound, and Churchill wanted to have them equipped with bridging devices to enable them to cross infantry trenches and pass along stretches of road which had been torn up by the Germans. On his own initiative he proposed the construction of steam-driven vehicles on the Holt-Caterpillar system, which could carry armoured protection, machine-guns and the necessary crew. The Director of Fortifications and Works was won over to the scheme, and so it was that support for the new weapon gradually widened.

Meanwhile the British offensive at Neuve Chapelle and la Bassée had been wrecked by barbed wire and machine-gun fire. The response was to assemble masses of troops, guns and ammunition for yet further offensives. The British, in other words, were going to fight like the Dervishes at Omdurman. To borrow the expression of von Schlieffen, it was a combat of 'the man with his bayonet against the flying bullet, of the target against the marksman' ('Cannae', in *Vierteljahrshefte für Truppenführung und Heereskunde*, 1910, 205). In the event both sides resorted to setting up broader belts of obstacles, and digging deeper trenches and dugouts; more and more the contest resembled a siege, and on long stretches of the front the combatants tunnelled and fought underground.

At the beginning of June 1915 the then Lieutenant-Colonel Ernest D. Swinton, R.E., laid before the British high command a paper on his machine-gun destroyer and how it should be used. In his turn Field Marshal Sir John French passed on the proposal to the War Office. This paper already contained in outline the essential technical and tactical specifications for the later prototype, and it emphasized in particular the importance of security, and the need to gain surprise through a full-scale attack: 'These machines should be built at home secretly and their existence should not be disclosed until all are ready. There should be no preliminary efforts made with a few machines, the result of which would give the scheme away.' (*Eyewitness*, p. 131.)

In February 1915, after an unsuccessful experiment in obstacle-crossing by a laden Holt Tractor, the British War Office had dismissed the idea of building 'landships'. The precise wording was that the project was 'out of the question' – a phrase which is familiar to us as well. But now at last the War Office was spurred into action by Swinton's memorandum and the knowledge that the Royal Navy was continuing its efforts, and the result was that it undertook further development in association with the Navy and the newly created Ministry of Munitions.

In September 1915 an experimental machine called 'Little Willie' underwent an unsuccessful test. But the device had not been built according to the latest specifications as laid down by Swinton, and more promising was a wooden mock-up of a new machine which was available for inspection at the same time. This was 'Mother', the later Mark I Tank, which made its first appearance at the front exactly one year later. Lieutenant W. G. Wilson, R.N., has described how the vehicle was built by W. Foster & Co., and how it assumed the characteristic rhomboid form, with the upturned nose and the track leading right round the hull. Firing trials were carried out with German machine-guns and ammunition against steel plate. Likewise an experimental training ground was selected and fitted out, complete with obstacles which corresponded to the German defences. The first driving and live firing trials were carried out as early as January 1916. The British had captured some German 50mm guns which were mounted in armoured cupolas, and there was concern that the Germans might introduce small-calibre guns with armour-piercing capability, which would have added greatly to the effectiveness of defence against tanks. Appropriate countermeasures were therefore taken into consideration. For the manpower base of the new armoured force, the British turned to the existing naval armoured car squadron. Finally the name 'tank' – soon to be known throughout the world – was chosen as cover for the new weapon.

On 2 February 1916 the first tank went through its paces before an audience of dignitaries including Lord Kitchener, Mr. Balfour and Mr. Lloyd George. The civilian ministers were full of enthusiasm, but Lord Kitchener was sceptical. He refused to believe that the war could be won by machines that would so easily be knocked out by the enemy's artillery. This was at variance with the opinion of a number of officers from the front, who favoured the new machine.

In the same month the indefatigable Swinton completed a memorandum on the future employment of the tank. It is still worth reading today because of its clear-headedness and what it has to say about the way armour is likely to be going. We accordingly quote a number of passages:[3]

'Since the chance of success of an attack by tanks lies almost entirely in its novelty and in the element of surprise, it is obvious that no repetition of it will have the same opportunity of succeeding as the first unexpected effort. It follows, therefore, that these machines *should not be used in driblets* (for instance, as they may be produced), but that the fact of their existence should be kept as secret as possible until the whole are ready to be launched, together with the infantry assault, in one great combined operation. The extent to which the attack is pressed, i.e., whether it is to be a step-by-step operation in which, after artillery preparation, a strictly limited advance is made over the front concerned and the gain of ground consolidated, and then, after the necessary pause to give time for a renewed artillery preparation of the enemy's new front line a further limited advance is made, and so on; or whether a violent effort is to be made to burst right through the enemy's defensive zone in one great rush, depends on the decision of the Commander-in-Chief and the strategic needs of the situation. But, so far as is known, a step-by-step advance – which has the drawback of giving the enemy time to reinforce the sector threatened – is not a course recommended for any positive advantages which it possesses. It is a course which has been forced on us by the inability, with the means hitherto at our disposal, of infantry even after an immense sacrifice of life to force their way through successive lines of defence guarded by machine-guns and wire, of which none but the first can be thoroughly battered by our artillery.'

'Not only, however, does it seem that the tanks will confer the power to force successive comparatively unbattered defensive lines, but, as has been explained, the more speedy and uninterrupted their advance the greater the chance of their surviving sufficiently long to do this. It is possible, therefore, that an effort to break right through the enemy's defensive zone in one day may now be contemplated as a feasible operation.' (*Eyewitness*, 203-4, 210.)

Swinton declares that in favourable terrain a daily average of twelve miles' progress is feasible. He sets the capture of the enemy artillery as his objective, and since the batteries will be deployed over a wide area the plan of attack must provide for a war of movement that is capable of sweeping them up. He rightly identifies the artillery as the most dangerous enemy of his new weapon, and states that the guns must be suppressed by one's own artillery and aircraft. The employment of gas and smoke has been discussed already.

It was fortunate for the Germans that from the outset the British shrank from following these guidelines. After the successful experiments and demonstrations the command of the British army in France put in an initial order for only forty tanks. Swinton protested, and got the War Office to order one hundred machines all in one go. These were to be produced by the Ministry of Munitions.

At the turn of 1915/16 the new arm was constituted as the 'Heavy Section of the Motor Machine-Gun Service' at Siberia Camp, Bisley, under the command of Swinton, who now held the temporary rank of colonel. The first complement of personnel was assigned at the beginning of March – the officers and men had had some training on machine-guns, and most of them appear to have had a first-class grounding in the technicalities of powered vehicles. Lieutenant Stern, R.N.V.R., and Lieutenant Wilson, R.N., who had already worked on the development of the tank, were taken on the establishment as majors.

In April the order for the tanks under construction was increased to 150, of which 75 were to be equipped with two guns and three machine-guns each, and the remaining 75 with machine-guns only; they were dubbed respectively 'male' and 'female' tanks. In addition to shells the tank gun was to fire canister rounds for close-range combat.

The new arm of service was organized initially in six companies of twenty-five tanks each. Before the first tank rolled, however, the new commander in France, Sir Douglas Haig, demanded tanks for his planned offensive on the Somme. This represented a very great danger of feeding the novel weapon into action by penny packets before it was ripe, so giving away the element of surprise.[4]

Meanwhile the work of setting up the tank forces went ahead. Among other measures Captain Martel, R.E. (of later fame) was given the task of creating a training area at Elveden in Suffolk. In the course of six weeks' work three battalions of engineers proceeded to lay out a replica of a sector of the Somme battlefield. It was more than 1½ miles in width, and in depth it included the British front and support lines, no man's land, and the first, second and third German lines, complete with obstacles, shell craters and so on.

A transmitter with a range of about three miles was tested to explore the practicability of wireless communication, and an unsuccessful experiment was made in communicating with aircraft by signal lamps. Communication between the tanks was to be carried out by metal discs and small flags waved out of the manhole in the roof. The Navy supervised the installation of de-magnetised compasses so that the tanks could stay on course.

From the beginning of June the fully equipped tanks arrived at Elveden and training could proceed. While this work was in progress, the British high command resorted to the old and usually ineffectual tactic of charging head-on against barbed wire and machine-guns. The great six-corps offensive made negligible gains despite an unprecedented employment of artillery.

At the end of June the British tank forces received a first visit from Colonel J. B. Estienne, the creator of the French tank arm. He urged the British not to use their tanks before the French were ready with their own machines, in order to retain the element of surprise.

When the first experimental batch of 150 tanks had been completed, the question arose as to whether to order a further run, and so avoid the many disadvantages of a break in production. However the British high command wished to acquire experience with a limited number of tanks on the battlefield before any new orders were placed. Still more urgent was the desire to score some success on the Somme, where the battle so far had bought tiny gains at an immense price, creating an unfavourable impression. In the middle of August a half company went to the front, and the other half followed later – which meant that the process of fragmentation had begun. Shortly afterwards the British high command forbade the further installation of wireless in tanks, because of the possibility of interference with existing stations; likewise the use of kite-balloons, to fly signals for the tanks, was forbidden lest they draw fire. Altogether a great deal was done to render the direction of the new tank forces more difficult, and very little to help it along.

The first of the tank companies that arrived behind the front line in August 1916 had first to put on a series of displays to satisfy the curiosity of various visitors, which brought with it the danger of wearing out the machines prematurely. The second tank company arrived in France only two days before it had to go into action; half its personnel had had only one day's training in live firing. The third company arrived in France on 14 September, but before it could reach the front the first two companies attacked on the Somme on the next day! The Battle of the Somme had been in progress for ten weeks, and yet the attempt was now made to breathe new life into it with just thirty-two tanks. All the same the first tank action in history resulted in the placing of an order for a further 1,000 machines.[5]

Although the first true mass production of tanks had now got under way, one of the most distinguished of the pioneers, Colonel Swinton, was left out in the cold. Colonel Hugh Elles, R.E., was entrusted with the command of the tanks at the front, while the setting up and training of the new units was handed over to a former brigade commander from the infantry.[6]

2. IN FRANCE

In France, just as in Britain, only a handful of individuals had addressed themselves to one of the most burning questions of the war, namely what is meant by 'shock'? (This section makes extensive use of von Heigl's *Die schweren französischen Tanks; die italienischen Tanks*, Berlin, 1925). They concluded that there was no future in simply using the existing weapons in ever-increasing quantities.

Quite independently of the British, the French too very rapidly hit on the notion of employing some kind of powered vehicle to overcome barbed wire obstacles. Working with Major Boissin, Deputy J. L. Breton of the French National Assembly built a four-ton wire-cutting tractor, which was tried out with a fair measure of success on 22 July 1916. The Technical Section of the Engineers then tried to convert the Filtz Tractor, which was a 45hp agricultural vehicle, to a machine-gun carrier. Ten of the machines were put through practical tests in August 1915, but their cross-country performance was disappointing.

The French efforts began to bear fruit only at the beginning of August 1916, when the then Colonel Estienne, commander of the artillery of 6th Division, saw how the British were using tracked vehicles, namely the Holt Caterpillars already mentioned, to haul heavy artillery. This was enough to persuade him to press ahead with the development of an armoured vehicle running on continuous tracks.

After two fruitless letters, Colonel Estienne submitted a third paper to the commander-in-chief, General Joffre. We quote his own words:

'Twice within a year I have had the honour of drawing Your Excellency's attention to the advantages of mobile armoured vehicles which are designed to facilitate the advance of the infantry. In the course of the last offensive I became more and more convinced how valuable such a kind of co-operation would be. I have completed a fresh and thorough analysis of the technical and tactical problems of developing a suitable vehicle. It will have a speed of more than six kilometres per hour, and will be capable of assisting the passages of obstacles by our infantry, even when they are laden with rifles and packs, and by our artillery.'

The outcome was that on 12 December 1915 Estienne was received by Joffre's chief of staff, General Janin. In his presentation Estienne explained how important it was for a large number of armoured vehicles to go into action at the same time. This, he said, was the only way to guarantee complete surprise.

Estienne was given leave to go to Paris to solicit support among the authorities, and especially at the Ministry of Defence, and to find an industrial concern which would be willing to take on the responsibility of construction, with all its risks. Renault's first response was negative, but Estienne contacted M. Brillié who was one of the engineers at the Schneider works, and he was able to convince him how urgent the task was. Brillié was all the more disposed to take over the project since Schneiders were already experimenting with the Holt Tractor. The works' chief engineer, Deloule, and the director Courville lent a hand, and within a matter of days the team had created a design which was suitable for mass production. Some delays were occasioned by new experiments with the Holt Tractor by the *Direction des Services Automobiles*, but in January 1916 Estienne was able to gain an interview with Joffre and win him over for the project, with the result that the French high command placed an order for four hundred vehicles.

Significantly, however, Estienne was pushed to one side by the technical authorities. He resumed his old command on the Verdun front, where he was tied down for several months.

The War Ministry proceeded to award the contract for a second batch of four hundred tanks to Schneiders' competitors, the Saint-Chamond works, where the celebrated Lieutenant-Colonel Rimailho was in charge of the development of the design. The resulting Saint-Chamond vehicle was considerably larger and almost double the weight of the Schneider tank. It had a forward-projecting field gun and a secondary armament of four machine-guns.

In the middle of June 1916 the French high command learned that the British were also at work building tanks. It now called Estienne to mind, and commissioned him to go to England to see how the Allies were progressing. As already mentioned, Estienne immediately sought to convey the importance of preserving surprise, and of holding back this new weapon until it could be sent into action by the French and British simultaneously and *en masse*. After he returned from England he set to work on a plan for a gigantic offensive by the tanks of both armies, very much as was actually put into effect in 1918. However the British did not have the nerve to wait until the French had caught up with them.

When the first tanks were almost ready Estienne was appointed commander of the new *Artillerie d'assaut* under the authority of the *Direction des Services Automobiles*. Although in the meantime he had been promoted to general, there was a wide feeling that he was an ill-used person who had already been 'put out to grass'.

On 15 August the first troop assembled at Fort Trou d'Enfer at Marly-le-Roi. It consisted of alarmingly juvenile officers who had only just passed out of Fontainebleau, and of equally inexperienced men, many of whom had never seen powered vehicles, and had first to be trained as drivers in the schools at Châlons and Rupt. September saw the arrival of the first Schneider tanks and the first Saint-Chamonds, and work could now begin. It soon proved necessary to open a second and then a third training centre – at Cercottes near Orléans, and Champlien at the southern edge of the forest of Compiègne.

Estienne divided the vehicles into 'batteries' of four tanks each; four batteries formed a 'group' under a captain or major; several such groups consituted a 'Groupement'. The first group of Schneider tanks came into being in December 1916, and the second in January 1917.

The French now had to address themselves to the solving of the many technical difficulties which had accumulated as a result of the sheer speed with which the development had proceeded. It also became clear that the original thickness of armour was proof against the ordinary German S round, but not the armour-piercing SmK. It was hardly surprising that the original delivery dates could not be met as promised, and particular difficulties were experienced with the heavy Saint-Chamonds, whose tracks were too narrow and exerted an excessive ground pressure, causing the tanks to dig into soft ground and stick fast. The result was that the spring offensive of 1917 had to be carried out exclusively with Schneider tanks.

It was only now, just before the first operational use of tanks, that General Estienne had to recognize that the two existing types of French vehicle were far too cumbersome. He set about designing a lighter, faster-moving tank which would weigh five or six tons at the most, and carry a machine-gun or a light gun. In the summer of 1916 he presented himself once more to Renault, and this time he was able to win them over to his project. As early as March 1917 Renault was able to demonstrate their celebrated and magnificently successful model, and in May they received an order for 1,150 tanks, of which 650 were to be armed with 37mm guns and the remainder with machine-guns. In October at Estienne's urging the order was increased still further to 3,500 machines, and divided between the firms of Renault (1,850 vehicles), Berliet (800), Schneider (600) and Delaunay-Belleville (280), while the Americans undertook to build a further 1,200 tanks. Two hundred wireless tanks were ordered in addition. Unlike the medium tanks, the *Chars légers* were divided into companies of three platoons of five vehicles each, making fifteen front-line and ten reserve tanks per company.

But we have run somewhat ahead of events. Even before the first third of the *Artillerie d'assaut* had been delivered the cry arose for the tanks to be thrown into action. The same call had already been heard on the British sector, and in both cases the commanders were unable to ignore it.

The first idea had been for the tanks to attack at Beauvraignes in March 1917, but this was abandoned after the Germans withdrew to the Hindenburg Line. So it was that the French tanks received their baptism of fire on the Aisne on 16 April 1917.[7]

3. First Deployments, Mistakes and Misgivings

See Sketch Maps 6 and 7.

We have examined some developments of interest behind the Allied lines. We now turn to the struggle on the Western Front. On the basis of their experiences in the 'Winter Battle', the French devoted weeks of meticulous preparation to their forthcoming 'Autumn Battle' in Champagne. The British were working on similar lines in Artois. The essential differences from the earlier pushes consisted in a considerable reinforcement of the artillery, a huge increase in its ammunition, a long-drawn-out period of artillery preparation, and the extension of the artillery targets deep into the enemy rear. The fire was to be directed with the help of a great number of observation aircraft.

The drum fire began on 22 September and the attack followed on the 25th. German guns numbered only 1,823 as against the 4,085 of the French; in Champagne six German divisions faced eighteen French, and in Artois twelve German divisions confronted twenty-seven French and British divisions. These were just the front-line forces, and we have to remember that the enemy had strong reserves, and the Germans very few.

The enemy put down a gigantic barrage (including gas shells in Champagne, and the British attack was supported by cylinder gas). The infantry assault followed. The enemy achieved a number of penetrations in both sectors, which in Champagne attained a depth of between three and four kilometres between Tahure and the Navarin-Ferme, and up to 3½ kilometres in Artois. There were times when the Germans were severely stretched because of the painful lack of reserves, but on neither sector did the Allies achieve their intended breakthrough. The offensive was prolonged by a series of mostly local actions which in Artois lasted until 13 October, and in Champagne until the following day. Fighting on the defensive, the Germans consumed 3,395,000 shells, and lost 2,800 officers and 130,000 men. The enemy fired 5,457,000 shells, which figure includes the British preparatory bombardment, but not the ammunition which the British expended during the battle. Allied losses amounted to 247,000 men, a sacrifice which was completely out of proportion to the ground gained.

From these battles the Allies drew the tactical lesson that: 'In a future attempt at breakthrough we should strive to win through a succession of combats, and not through an attack in a single bound.' (Reichsarchiv, IX,101.) In addition it was thought that the quantity of artillery and the supply of ammunition must be increased still further, but no very clear conclusions emerged about the effectiveness of gas.

The idea of conducting the offensive step by step – of breaking it down into individual actions – only played into the hands of the Germans. Large-scale surprise was now out of the question, and the defenders had time to make ready their reserves behind the threatened sectors, and build additional positions to the rear. The Allies tried to make a virtue out of necessity, and persuaded themselves that they had hit on a way of wearing down the enemy reserves little by little until they could finally break through the weakened German front. The artillery battle meanwhile degenerated into a war of attrition.

The German high command adopted essentially the same way of thinking, and they applied it at Verdun, after they had failed to take it by surprise in the spring of 1916. 'The decision to take the fortress of Verdun by an accelerated attack is based on the well-established efficacy of heavy and super-heavy artillery. To this end we must choose the most suitable sector for the attack, and conduct the artillery strike in such a way that the infantry breakthrough is bound to succeed.' The initial attack, with its 'crushing force', was confined to the east bank of the Meuse, and even there it was limited to the north-eastern corner of the Côtes Lorraines. (Reichsarchiv, X,58.)

Twelve hundred guns stood ready together with considerable stocks of ammunition. The aim was to gain the objective 'in the first onrush', and it was emphasized that the attack must 'on no account be allowed to bog down, so as to prevent the French from re-establishing themselves in rearward positions, or reorganizing their defences once we have broken through'. In the event, however, the offensive was almost immediately converted to a series of step-by-step pushes – a howling contradiction which was immediately evident to the troops who were responsible for putting the thing into effect. It was solely due to the offensive spirit of the troops that they managed to push beyond their objectives and score a number of successes which actually exceeded the expectations of the high command. Such an episode was the capture of Fort Douaumont on 25 February, the fifth day of the offensive. This fortification was stormed upon the independent initiative of three officers who chose to ignore their assigned sector and the objective which had been set for that day. The officers in question deserve to be named: Captain Haupt, First Lieutenant von Brandis and Reserve Lieutenant Radtke of 24th Brandenburg Infantry Regiment.

This was the high point of the 'accelerated attack', which now gave way to a battle of attrition. There were no reserves available behind III Brandenburg Corps to exploit its success. On 26 February the high command turned down Fifth Army's request for reinforcements to enable it to extend the attack to the west bank of the Meuse. By the 27th there were signs of exhaustion among the attacking troops, enemy resistance was stiffening, and casualties were rising. In the seven days of fighting since the start of the offensive the Germans had lost 25,000 men, while taking eight kilometres of ground, 17,000 prisoners and eighty-three guns, but from now on gains would be made only step by step and at disproportionate cost. The offensive was extended to the west bank of the Meuse at the beginning of March, and an intensive use was made of gas in an attack at Fleury on 23 June, but again no decisive successes were achieved.

After the battle of attrition at Verdun had been prolonged with extraordinary obstinacy for four months, the enemy opened an offensive on their own account on the Somme. It was delivered with unprecedented force, as we shall see. Meanwhile the wasting battle at Verdun had made frightful inroads on the hitherto intact core of the German infantry, and destroyed the confidence of the troops in their leadership. By the end of the fighting no less than forty-seven German divisions had been engaged at Verdun, six of them twice over, and the Germans had fired off 14,000,000 rounds of artillery ammunition, and taken 62,000 prisoners and two hundred guns. Over the same period the French had committed seventy divisions, thirteen of them twice over, and ten of them three times. The imbalance of forces was all the greater since the French divisions comprised four regiments of infantry, and most of the German divisions had only three. In dead, wounded and missing the Germans lost 282,000 men, and the French 317,000. While the attack on Verdun tied down the forces available to the Germans on the Western Front, it left the offensive potential of the British completely intact, and weakened the French potential only in part, and certainly not to the extent that was needed to disrupt the long-meditated enemy attack on both banks of the Somme.

On 1 July 1916 the Anglo-French offensive struck the 12½ divisions of the German Second Army. A preparatory bombardment from 3,000 guns had been raging from 24 June, and now seventeen divisions attacked in the first wave, with fourteen infantry and three cavalry divisions following in reserve. The Allies had 309 aircraft available for the battle, which gave them the mastery of the skies, and the 'German air defence was confined to immediate close reconnaissance' (Reichsarchiv, X,347). The Germans had at their disposal only 104 aircraft and 844 guns.

Dust, smoke and the morning mist covered the preparations of the enemy forces until they burst forth at 0830, and by the end of the first day of the offensive they had won the forward German trenches on a frontage of some twenty kilometres and to a depth of up to 2½ kilometres. These gains were expanded on the following night. There was a slackening in the attacks from 3 July until they gradually increased again in force, which induced German Second Army to call for the setting up of special machine-gun companies and combined machine-gun/sharpshooter units. These worked extremely well and their intervention often proved decisive.

On 14 July the battles flared up again with the launching of a major new offensive; its gains, however, were small, and some of them were lost again when the Germans counter-attacked on the 18th. The Allies followed up with a further offensive on 20 July which pitted sixteen divisions against eight German; this too was beaten off. After violent local combats the enemy put in a powerful attack north of the Somme which was rewarded with insignificant gains. The battle roared on with large-scale Allied pushes on 7 August, 16-18 August and 24 August, but these too were in vain.

By this time the Allies had lost 270,000 men and the Germans 200,000. The enemy had broken into the German line over an area measuring twenty-five kilometres wide and at most eight kilometres deep, but there was no question of an actual breakthough. Altogether the fighting had ranged 106 enemy divisions against 57½ German divisions.

The endurance of the British infantry had been severely tested in the process, and public opinion at home was shaken. The British high command took due note and concluded that it could justify fresh attacks only with the support of new weapons. The command postponed the resumption of the attack until September, and decided to commit the first of its newly arrived companies of tanks.

Thus the first thirty-two tanks rolled forward to the attack through the morning mist on 15 September. This was not a particularly large number, and the concentration was diluted still further by the way the machines were divided between General Rawlinsons's Fourth Army and General Gough's Reserve Army. In addition a number of tanks were inevitably lost through mechanical failure. All of this was deplorable, and yet the appearance of these few machines enabled the British to score their greatest success so far. Such was the impact of this novel weapon. The offensive spirit of the British infantry revived at once, as witness the celebrated message from an aircraft: 'A tank is walking up the high street of Flers, with the British Army cheering behind it.' Public opinion also responded to the good news from the front line. Naturally, however, a handful of tanks could not aim to break through a German position which had been consolidated over the course of ten weeks of heavy fighting; there were not nearly enough machines for that purpose.

After this first engagement Colonel Elles was given field command of the British Tank Corps; he held this post until the end of the war and made a significant contribution to the development of this new arm of service. The British high command now asked for the construction of 1,000 tanks.[8]

On 25 and 26 September thirteen tanks rolled against Thiepval over ground that was boggy and pock-marked with shell holes; nine became stuck in the craters, two more broke down, and only two tanks actually reached the village. One of these machines, supported by a single aircraft, nevertheless took more than 1,000 metres of trench and captured eight officers and 362 men; in less than one hour the British infantry were able to secure the gain with the loss of only five men.

But the action at Thiepval, like all the rest that followed in the course of the autumn, was carried out by a small unit of tanks. A very large number of machines was available, and yet no attempt was made to use this force against a single objective. While all the other arms of the service, including aircraft, were being concentrated on the battlefield in ever greater numbers, the British high command did exactly the opposite with the tanks, even though it had originally approved what Swinton had to say about concerted action. Swinton wrote with every justification: 'With the example before us of the stupendous mistake of the Germans in first releasing gas over a short sector, we, sixteen months later, with our eyes open, committed a similar error. We threw away a surprise.' (*Eyewitness*, 297.) This corresponds to a statement in one of the British official accounts, which speaks of the sheer wastefulness of forfeiting surprise in such a way, and again makes the comparison with the German use of gas at Second Ypres.

Now the cat was out of the bag. The French, General Estienne in particular, were beside themselves with fury, believing that they would now have to reckon with stronger German countermeasures, and conceivably also with German tanks. But here they gave too much credit to the Germans. Our high command admittedly called for experimental work on tanks, and offered a reward for the recovery of the first British machine, but for the moment that was all. Anti-tank ammunition for the infantry was neither requested not developed, and on 17 November the army group commander Crown Prince Rupprecht issued the following order of the day: 'Infantry can do little against tanks by themselves, but they must nevertheless be schooled to believe that they can hold out in full confidence that the artillery will intervene to exorcise the danger.' In other words the moral element was supposed to beat the material, at least as far as the German infantry were concerned!

The artillery did take countermeasures of a sort: twelve batteries of field artillery were reformed for this purpose, and five batteries of six field guns each were set up for close-range combat. They were to be deployed immediately behind the front line, to engage the tanks with armour-piercing shells. The engineers too were busy. They dug ditches and tank traps, laid out minefields in suitable locations, and converted areas of ground to swamps by damming streams. Finally the Minenwerfer were equipped with special carriages for low-trajectory fire.

When the summer battles were over the French and British prepared a great blow for the spring of 1917. They planned to pin down the greater part of the German reserves by a Franco-British attack at Arras, then break through the German front in the Champagne hills and between Reims and the Chemin-des-Dames, and finally exploit the breakthrough with powerful reserves. The British tanks were to support the attack at Arras, while the French machines did the same at Berry-au-Bac on the Aisne.

The sixty available British machines were split up among the various corps for the battle at Arras on 9 April. They performed some useful local services, but they were too dispersed to secure a large-scale success. However the British established that the Germans had no serious anti-tank defence, apart from the armour-piercing rounds for their close-combat artillery. The Germans on their side captured their first British tank, evidently one from the original batch. They performed tests against the machine's armour, and found that the only effective weapons available to their infantry were the SmK round, special cartridge loadings and Minenwerfer firing at low trajectory.

On 16 April 1917 the French tanks rolled forward to their baptism of fire at Berry-au-Bac. They were deployed on Fifth Army's sector in two groups which had the mission of breaking through the German front in a single bound and within a span of twenty-four or at most forty-eight hours, and then rolling up the defences to the east. The ground rose gradually towards the German positions, but offered no particular difficulties. The area was bordered to the east by the Aisne, to the west by the heights of Craonne, and divided down the centre by the valley of the Aisne, a stream which was three metres wide and bordered by water meadows and bushes. The battlefield was overlooked from the north-east and north by the height of Prouvais, the high ground south of Amifontaine. The only obstacles in the way of the tanks were the trenches of the two sides and the shell holes, and even these scarcely existed north of the Corbény–Guignicourt road. The main danger was presented by the commanding observation posts available to the German artillery.

The offensive was preceded by a fourteen-day bombardment from 5,350 guns, which left the Germans in no doubt as to the width and objectives of the attack, and enabled them to arrange their defences in depth and bring up fresh forces – notably in the form of guns, and reserves of infantry which were held ready for counter-attacks. On 4 April, indeed, a successful counter-stroke by the German 10th Reserve Division at le Soldat, south-east of Berry-au-Bac, captured 900 men together with a number of orders concerning the intended attack. Altogether on this single day the French had to bring down defensive fire from 250 new battery positions. In addition the German artillery was quicker off the mark than before, and had greater material resources at its disposal. In fact the purposeful way that Hindenburg and Ludendorff had fought their battles on the Eastern Front, as evident in the 'Siegfried Manoeuvre', was now harvesting its first major success in the west. This time there could be no question of the Germans being taken by surprise.

For their offensive the French put together sixteen infantry divisions, two brigades of Russian infantry, and a cavalry division. The attack was supported by 3,800 guns, more than 1,500 mortars and by no fewer than 128 Schneider tanks – by far the most powerful concentration of armour so far seen on any battlefield. The most important directives for the French tank attack ran as follows (this section makes use of Lieutenant-Colonel Ferré's 'Le premier engagement des chars français', in *Revue d'Infanterie*, 1 April 1936):

'The tanks will accompany the infantry attack, opening the way though barbed wire obstacles and covering the advance.

'The tanks are armed with artillery pieces and machine-guns, but their most potent weapon is simply to keep on going. They open fire at close range – 200 metres at most with the gun, and 300 with the machine-gun; at longer range they will fire only in exceptional circumstances.

'Tanks and infantry remain in close association during the combat, but the tanks do not wait for the infantry if they see an opportunity for going forward; once the attack has opened the tanks advance on their objectives, and halt only when they encounter obstacles which they cannot cross with the equipment available to them. When our infantry catch up with a tank which is detained in this way, they must do everything possible to help it to overcome the obstruction. If the infantry are held up by enemy resistance before our tanks are on the scene, they must lie down and wait for the tanks to intervene. The tanks will roll through them and on against the enemy, suppressing the hostile fire. In such a way the tanks and infantry will lend mutual support as they advance towards their common objective; they wait for one another only when their own resources do not allow them to progress any further.'

A supplementary order was issued on 23 March, the purpose of which was to eliminate ambiguities in the instructions just mentioned. This made it clear that the tanks were supposed to fall in with the tactics of the infantry.

Sixteen tanks made up a 'group'. Five groups of this kind fought as the 'Groupement Bossus' to the east of the Miette on the sector of XXXII Corps, while a further three were designated the 'Groupement Chaubès' to attack on the sector of V Corps, west of la Ville-aux-Bois and the Miette. The initial advance was to be made in columns, upon which the tanks were to deploy in line for combat with intervals of 45-50 metres between the machines. A special infantry company was assigned to each group, to help with overcoming obstacles and in close-range combat. The attack was to follow a creeping barrage, which was to lift 100 metres every five minutes.

The tanks were to intervene only when the offensive reached the third or fourth German lines, to make up for the falling away of artillery, and to help the infantry forward. This meant that they were go into action four hours after the opening of XXXII Corps' attack, and 3½ hours after that of V Corps. On the day before the offensive the tanks were to assemble in the area west and south-west of Cuiry-lès-Chaudardes, and on the following night the Groupement Bossus was to put itself in a state of readiness south-west of Pontavert, and Groupement Chaubès in the wood south-west of Craonne. From there the two forces were to set out respectively thirty and twenty minutes after the attack began. We specify the details from the orders:

Groupement Bossus was to move in a single column by way of Pontavert to le Choléra, where it would split into two columns – the left-hand column, consisting of the leading three groups, to continue the advance between the le Choléra–Guignicourt road and the Miette, while the right-hand column, comprising the two rearwards groups, pushed initially along the le Choléra–Guignicourt road and then, after crossing the first German line, headed in the general direction of the Prouvais height. The approach was to be made in single file, and only after the first German line had been passed was the column to deploy into combat formation. The tanks had to wait for the barrage to cease, meanwhile occupying the line they had reached four to five hours after the offensive opened. Only then would they attack the third German line, and press on to Guignicourt and Prouvais; last of all, the three left-hand groups would attack Provisieux. The objectives were subdivided meticulously among the groups; after the attack they were to reassemble north-west of Guignicourt. Repair teams came up behind each group, and every column was in turn followed by a recovery group. Groupement Chaubès would likewise move in a single column, north-westwards by way of le Temple Ferme towards Amifontaine. Once the first German line had been crossed it was to form two columns, and then after crossing the second line deploy into combat formation and proceed to the attack. After the attack it would assemble west of Amifontaine.

Such at least were the plans of the high command. In the event the mighty preparatory bombardment damaged only the first and second German lines, and these only in part; the rearward defences were virtually unscathed. On 16 April the tanks were standing in readiness at the appointed time; Groupement Bossus was up to strength, but Groupement Chaubès had lost eight tanks which had bogged down. As was stipulated in the plan of attack, the infantry alone followed up behind the barrage. They took the first German line with no great difficulty, but it was a different story at the second line. The French arrived here between 1000 and 1100 only after they had suffered severe losses in heavy fighting, and their gains were confined to a sector running from Caesar's Camp by way of Mauchamps Ferme to the Old Mill south of Juvincourt; from here the line described a re-entrant to la Ville-aux-Bois, which the French were unable to capture. Further to the west the attack made no significant progress beyond the first German line, and at Craonne it failed altogether.

Meanwhile Groupement Bossus set off at 0630 in a column measuring two kilometres long. Progress was slow, because the road was encumbered with infantry and artillery. At 0800 the head of the column reached the bridge over the Miette west of le Choléra; it was under heavy German artillery fire, but only one machine was hit. Two tanks broke down completely, and two more got on the move again after being repaired. The escorting infantry had prepared the passage over the French front line, which was passed without difficulty, but the German front line occasioned a delay of forty-five minutes, and the first tank did not reach le Choléra Ferme until 1015. Here the supporting infantry had already been broken up by enemy artillery fire, and they lost contact with the tanks. At 1000 the western, or left-hand group deployed into the attack. Immediately Major Bossus's tank received a direct hit on the top of the machine which killed the occupants and set the tank on fire, depriving the attack of leadership at the decisive moment. We shall now take up the story of the individual groups.

A few minutes later the foremost group crossed the German front line, accompanied by weak units of the attacking infantry; over to the left there seemed to be French infantry advancing on Juvincourt, but nobody was coming up on the right. Seven tanks managed to cross the German trenches, but a further seven broke down. Shortly after 1200 the seven surviving machines reached Hill 78, and crossed the third German line behind, calling in vain on the infantry to follow. Two of the tanks were now put out of action, but the crews seized a German dressing station and took a number of prisoners. Two more tanks were shot up at 1315 and 1330 respectively. Finally the last three tanks made their way back in order to regain contact with the infantry, and they encountered nine tanks of the following group, the 6th, together with one of their own tanks which had been repaired.

The 6th Group had lost two tanks through breakdown when crossing the German second line. On the far side it attacked pockets of German troops which had been interfering with the advance of the French infantry, then lost five machines to artillery fire at a range of between 1,800 and 2,000 metres, and finally deployed just to the right of the remnant of the forward group, the 2nd. At 1430 the thirteen tanks of these two groups beat off a powerful German counter-attack against Hill 78. Captain Chanoine, their joint commander, decided against plunging on without support, since the infantry to the right had been making no progress, and he recrossed the line south of Hill 78 so as to withdraw out of reach of the enemy fire. A little later Chanoine was able to establish contact with the commander of 151st Regiment of Infantry, which had now reached the area between Mauchamps Ferme and the Miette, and they concerted an attack with the limited aim of recapturing Hill 78; the assault went ahead and the infantry reached the objective between 1730 and 1800. By agreement with the regimental commander the tanks now fell back along the Miette to le Choléra; in the attack they had lost one machine to artillery fire, and another four which were stuck in shell craters.

Group 5 delivered the third attack. It deployed to the right next to Group 6, waited for the artillery barrage to end and attacked at 1200. The French managed to take the third German line and occupy it with their supporting infantry. The attack was continued by nine tanks which drove through a copse to the north-east of the captured sector of the line, and reached the Guignicourt–Amifontaine railway line without having encountered any further obstacles or enemy fire. Here the French lost one tank to artillery fire and another to a breakdown. Meanwhile Group 5's commander had made contact with the commander of the following regiment of infantry, the 162nd, only to be told that the regiment had been too severely mauled to be able to continue the advance. Shortly after 1700 the tanks beat off a German counter-attack against the same regiment, and the machines were then withdrawn behind the infantry for the night.

Fourth in line came Group 9's attack. It was held up by other troops for a considerable time at le Choléra, but thirteen tanks reached Mauchamps Ferme and launched their attack from there at about 1300. They ran into artillery fire and were wiped out south-west of the railway; the infantry did not follow them into the attack.

Last of all Group 4 advanced in two columns between the Miette valley to the west and the valley of the Aisne to the east. It reunited on the second German line on the Aisne. After considerable delays and losses the five leading tanks opened an attack at 1500 in support of some infantry who were about 600 metres to the north and in a shaky condition. Two tanks were set on fire and the others returned after beating off a German counter-attack. The rest of the group pushed along the Aisne and was able to clear a trench at 1530; two tanks were put out of action, but the supporting infantry of 94th Regiment were able to follow up the success. The group was then withdrawn to its original assembly area.

Finally we turn to the three groups of Groupement Chaubès, which set out from their forming-up area at 0630. They advanced in a single column by way of le Temple Ferme, but were spotted by German aircraft, tracked by artillery observers, and came under concentrated fire. A number of delays were occasioned by the fact that the supporting infantry failed to make passages in good time across the French and German trenches. The commander's tank of the leading group was immobilized by a hit, and the following group ran into the vehicles from behind. The German artillery fire increased all the time, and the crews of the knocked-out tanks extricated their machine-guns and went into battle alongside the infantry, who were likewise pinned down. In the evening no more than nine of Groupement Chaubès' tanks were able to return to the assembly area under their own power; essentially the group had been destroyed at a range of between three and six kilometres by the indirect fire of one battery of field artillery, two batteries of heavy field howitzers, one battery of 10.0cm guns and two mortar batteries. Under this fire the tanks had been advancing at the pace of the supporting infantry.

Altogether the attack of 16 April 1917 proved to be a costly failure. The tank crews numbered 720 men, and 180 of them, or 25 per cent, were dead, wounded or missing. Of 121 tanks which set out from the assembly areas, 81 were lost – including 28 through breakdown, seventeen through artillery hits alone, and 35 through fires occasioned by artillery. A number of tanks caught fire without being hit at all. Twenty tanks were retrieved. The final loss came to 76 tanks out of a grand total of 132, or 57 per cent.

After the battle the French drew the following conclusions from their failure:

(a) the tanks had inadequate cross-country mobility;

(b) the supporting infantry rendered virtually no help;

(c) no tank had been disabled by small-arms fire, and the armour had met expectations;

(d) on the other hand 57 tanks were knocked out by artillery fire: fifteen through direct fire, and 37 through indirect fire – mostly from heavy-calibre

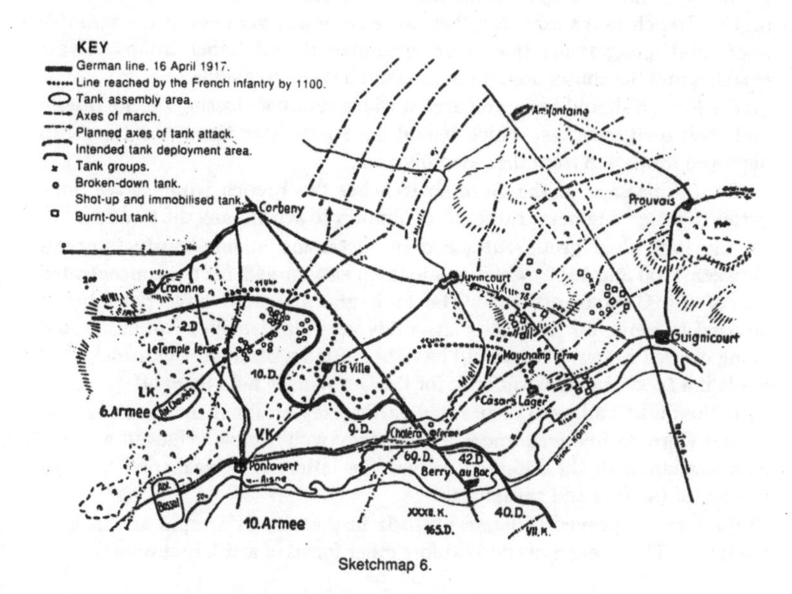

Sketchmap 6.

guns. This was due to the effectiveness of the German artillery observation, and the fact that on 16 April the German guns had been virtually untouched by the French artillery; in future something would have to be done to beat down the defenders' guns and blind their observation posts;

(e) most of the losses had been sustained when the tanks were moving in column, or during the holdups or deployments. Losses could be diminished by advancing in deployed formation from the assembly areas, and these in turn should be located close behind the infantry's start-line;

(f) the main reasons for the disappointing results of the tank attack were associated with the failure of the offensive as a whole, which left the tanks vulnerable. The infantry had been exhausted and decimated during the previous fighting, and they were in no condition to exploit the successes which the tanks had achieved in the general direction of Guignicourt and the height of Prouvais;

(g) tanks, even when operating individually, proved very effective against moving infantry, as was shown by the rapid collapse of the enemy counter-attack west of Guignicourt;

(h) against dug-in infantry, however, lasting success could be achieved only when the attacking infantry was in a position to exploit without delay; otherwise the gains of the tanks proved to be as costly as they were useless. The French drew the further conclusion that tanks must fight only in close assocation with infantry – a notion which dominates French tactics even now.

From the German perspective we must also add:

(a) the French tanks executed their long approach marches at the infantry pace, and along roads that were encumbered with other troops. These marches and the consequent holdups could have been avoided;

(b) the French should have prepared a whole series of crossings of the Miette and their own trenches, which would have permitted them to advance in deployed formation from the assembly area;

(c) for the passage of the German trenches the French would have done better to assign engineers rather than infantry to accompany the tanks;

(d) the individual groups attacked in succession at one-hourly intervals between 1100 and 1500, which made them easy targets for the concentrated fire of the German artillery. The tank groups would have been better advised to move up from their assembly areas in deployed formation and along cleared routes. This would have permitted a simultaneous attack, and made it a good deal more difficult for the Germans to hold them off;

(e) if the tanks had been committed earlier, say at about the time that the second German line was under attack, they would have achieved a closer co-operation with the infantry, before the latter were weakened through prolonged fighting and casualties;

(f) the barrage proved to be an obstacle in the way of a rapid advance by the tanks. There was a need to explore other forms of artillery support;

(g) for all the mistakes in the way the assault had been prepared and carried out, the tanks still managed to advance 2 to 2.5 kilometres further than the infantry. The foot soldiers were unable to follow, despite the weak resistance on the part of the Germans and the slowness of the tanks. We can only conclude that the main striking force of an offensive resides in tanks, and it is a question of developing the other arms in such a way that they can keep up with them;

(h) a complete breakthrough on 16 April 1917 would have been well within the bounds of possibility, if only the tanks had been employed in a more effective way, and if the offensive tactics of the other arms had been brought into full accord with the performance of the new arm, the tank.

At that time, in 1917, the Germans were impressed by their successes in beating off the attack, and they came to other conclusions. They should have entrusted the main responsibility for dealing with tanks to the direct fire batteries. Instead, the batteries in question were gradually disbanded, since the Germans believed that they had a complete anti-tank defence with the infantry firing SmK rounds and special cartridge loadings, and the artillery, especially the heavy calibres, shooting from long range.

In France the disappointment at the failure of the attack on 16 April 1917 led to some violent criticisms being levelled against the tanks. However it was not long before the worth of armour was vindicated in further combat, and indeed the contribution of the tanks had already been recognized in official circles, as witness the GHQ *Ordre général* No. 76 of 20 April that year:

'The tanks were the first of our forces to penetrate the second enemy line in front of Juvincourt, and they secured its conquest. This was their first appearance on the field of battle, and they won for themselves a place of honour alongside their soldier comrades, demonstrating what we may expect from the *char d'assaut* in the future.'

We turn now to the story of Groupement Lefèbvre, with its two groups of Schneider tanks and its single group of Saint-Chamonds. It had not been involved in the attack of 17 April in Champagne, but was committed a little later, on 5 and 6 May, in the fighting at Mannejean farm and the mill at Laffaux. These were elements in an attack by 158th Infantry Division, the composite Division Brécard and 3rd Colonial Division, which had the limited objective of pushing to the northern edge of the heights of the Chemin-des-Dames. Of the available tanks 158th Infantry Division was assigned the Saint-Chamond group and a 'battery' (four tanks) of Schneiders; one group of Schneiders went to Division Brécard and the remainder of the tanks stayed in reserve. On this occasion the splitting up of the tanks was justified by the conformation of the terrain. The missions of the groups and batteries were specified in detail, and 17th Light Infantry Battalion had been given a lengthy training in infantry-tank co-operation. The terrain on the plateau of Chemin-des-Dames was particularly suitable for tank operations because of the good going and the well-sited assembly areas on the southern slopes. It also presented difficulties to the Germans in the way

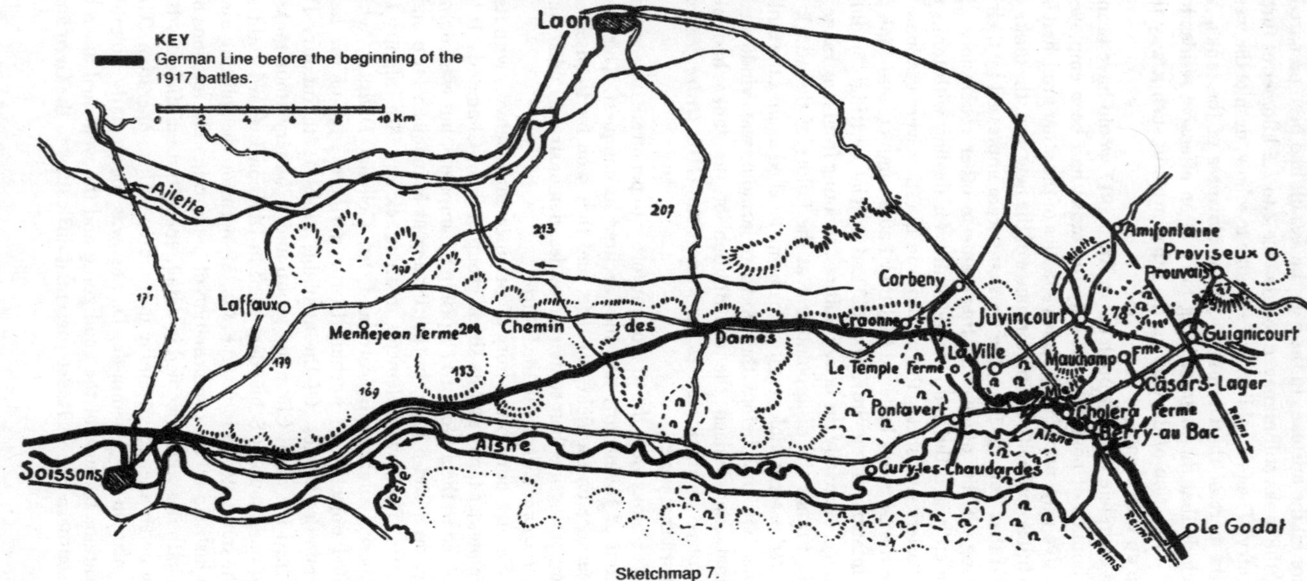
CHEMIN-DES-DAMES AND THE LASSAUX ANGLE
KEY
German Line before the beginning of the 1917 battles.
0 2 4 6 8 10 Km
Laon
Ailette
213
207
Laffaux
Mennejean Ferme
Chemin
des
Dames
Craonne
Corbeny
Juvincourt
Amifontaine
Proviseux
Prouvais
Guignicourt
Mauchamp Fme.
Casars-Lager
La Ville
Le Temple Ferme
Pontavert
Choléra Ferme
Berry-au Bac
Cury-les-Chaudardes
Le Godat
Reims
Aisne
Miette
Vesle
Soissons
171
179
169
193
198
78

Sketchmap 7.

of observation. There were unfavourable features however; these included the awkward approach routes, and a deep zone of craters after the French artillery had tried in vain to clear the belts of wire with delayed-action fuzes; the shell holes were to prove fatal to several of the tanks.

The subsequent attack was not particularly successful in itself, but the few gains that were made were due primarily to the tank forces, which lost considerably less in machines and personnel than on the Aisne. Moreover the high command and the other arms were pleased with what the tanks had done, which secured the future of the new weapon. Just as had been shown on the Aisne, it was clear that armoured attacks could gain lasting success only when they were followed up without delay by the infantry. Once again this had not occurred, even though the tanks had given the agreed signals, and on occasion even driven back to the infantry in an attempt to get the foot soldiers to occupy the positions that had been cleared.

The French scored a rather greater success on 23 October 1917, when they attacked the angle of the line at Laffaux. This time a considerably greater number of tanks was devoted to the assault. What was the wider background? The French had sustained heavy casualties during the fighting in the spring of 1917, and by the autumn had had no alternative but to await the entry of the Americans, meanwhile confining themselves to a series of small probes which were designed to improve the lie of the front line and try out new tactics. Moreover the French intended to carry out a number of augmentations by the summer of 1918 – increasing their heavy artillery two-fold, building 2,000–3,000 Renault tanks, and assembling large stocks of gas and smoke shells.

One of the small enterprises in question was to capture the Chemin-des-Dames. The operation was to begin with an attack on the Laffaux angle on a frontage of some eleven kilometres. There was no question of being able to take the enemy by surprise, and by the time the attack began the French had identified the presence of seven fresh German divisions and sixty-four new batteries. The German positions were solidly built, and the belts of barbed wire extended in some places to a depth of ten metres. The defending troops had cover in the numerous shelters and dugouts. A rearward line stretched along the north bank of the Ailette, and it lay beyond the objectives of the French attack. On the other hand the ridge of the Chemin-des-Dames descended steeply to the north, which among many sectors cramped the depth of the German positions and limited their fields of fire, compelling the artillery observers to post themselves in the front line.

The French carried out the attack with six divisions in the first wave, and six more in the second. The equipment and training of the troops had been fully restored since the spring battles, and in particular the French had rehearsed co-operation with tanks, and proved the ground with a number of probing attacks. The offensive was to be prepared by 1,850 guns firing some 3,000,000 rounds, while sixty-eight tanks executed the assault.

The tanks were organized into three groups of twelve Schneiders each, and two groups each of fourteen Saint-Chamonds, with a number of tanks in reserve. Each group was furnished with a combat supply element and a mobile workshop, while every *groupement* had at its disposal a repair and replacement train. At the end of August two squadrons of dismounted cuirassiers were assigned as close support infantry and trained in working with the tanks, and the main body of assault infantry exercised jointly with the armour. Aerial photographs underwent constant study, and the routes were reconnoitred and improved.

The offensive was ushered in by a six-day bombardment. At the points where the tanks were due to break in, the lanes through the obstacles were cleared by shells primed exclusively with instantaneous fuzes, which did not have the usual effect of creating deep of shell holes. Aircraft were ordered to report the progress of the attacking infantry and armour, and artillery spotter aircraft were assigned to keep an eye on the movement of enemy reserves and anti-tank artillery.

The tanks were distributed among five of the six assaulting divisions. Tank liaison officers were attached to the commanders of the infantry regiments, while the senior officers of the tank forces were located with the divisional commanders and the commanding generals.

A number of losses were incurred while the French were in the process of entering the assembly areas on the night before the attack. Half of Group 12, which was under the right-hand division, namely the 38th, was eliminated through breakdowns and German artillery fire; it was much the same story with Group 8, which was placed under 43rd Division; Group 11 with 13th Division reached its start-line without serious trouble, as did Group 31 (Saint-Chamonds) of 17th Division, and Group 33 (Saint-Chamonds) with 28th Division. However only 52 out of a total of 68 tanks reached the start-line, which shows how dangerous it was to keep tanks waiting within effective range of the enemy artillery, even though the Germans did not know the tanks were coming up and their artillery strikes amounted to no more than harassing fire on the roads during the night.

The tanks set off behind the infantry at 0505, while it was still dark, and they continued at the infantry pace. All the tanks of the right-hand group, the 12th, were *hors de combat* before they could reach the first objective. Group 8 was able to enter the fight with six of its tanks, though only after the French infantry had begun to advance on the second objective; the tanks moved forward in the space between the creeping barrage and the first wave of infantry, and a number of machines followed later, after they had undergone mechanical repairs. By 1100 eight tanks of this group had reached the objective, and covered the infantry while the troops were consolidating themselves there. Group 11 set off with 13th Division according to plan, and played a considerable part in its subsequent success, with twelve tanks reaching the objective. Group 31 did reasonably well, but Group 33 failed to reach the first German trenches.

On 25 October the French reached the Ailette without any further help from their tanks. By 1 November 1917 the Germans had evacuated the whole of the Chemin-des-Dames, and in addition to their casualties they had lost 12,000 prisoners and 200 guns. French losses amounted to 8,000 men, or 10 per cent of the total in action. Of the 68 tanks that had been engaged: nineteen were lost during the fighting, though only eight of them were lost to enemy action, the remainder became bogged down; twenty gained their objectives; five functioned as wireless tanks.

The casualties among the tank crews amounted to 82, or 9 per cent, which was of the same order as in the infantry. Most of the men had been hit when they were outside the tanks or had opened the hatches to get their bearings.

On this occasion the French deduced the following:
(a) tanks became effective against fortified positions only after they had crossed the zone of craters;
(b) flanking units were particularly vulnerable, and need special protection on this account;
(c) tank attacks must be carried out in depth. Objectives should be assigned not to individual machines, but always to whole units at a time, in other words to platoons or groups;
(d) the attempt to comnmunicate with infantry by flags had failed, and the only effective means was by word of mouth;
(e) the tanks were liable to heavy losses whenever they were standing still within sight of the enemy, and in future this should be demanded only in case of emergency;
(f) close co-operation with the infantry proved its worth – and indeed has remained the fundamental tenet of French tank tactics until the present day (Commandant Perré, 'Les Chars à la bataille de la Malmaison', in *Revue d'Infanterie*).

Concerning the last point we must add that being tied to the infantry as closely as they were on 23 October 1917, the tanks avoided being wiped out only because the Germans had no kind of anti-tank defence. The only effective anti-tank weapon available to them at that time was their artillery, and the unfavourable local conditions made it almost impossible to use; otherwise those huge and trundling targets would have shared the fate of the tanks on 16 April. In future such tactics will be suicidal.

So much for the first French tank battles. We return to the British, who had determined on a mighty offensive in Flanders, at whose ports German U-boats were based. There was no question of a surprise attack – quite the contrary! The intention was to gain ground step by step, and only after that ground had been comprehensively pounded by artillery, gassed, and if necessary blown up by subterranean mines. This was to be a battle of brute force and attrition, sedulously avoiding any kind of novel or untried technique – indeed deliberately renouncing the possibility of exploiting any unexpected success which might come the way of the British. Such was the thinking behind the Third Battle of Ypres.

On 7 June 1917 the British blew up the German positions on the Wytschaete salient, destroyed five German divisions and reached the Lys. This first blow secured the right flank of the coming offensive which was preceded by four weeks of bombardment and lasted until early December. The British tank forces were committed time and time again, but invariably in penny packets with strictly limited objectives, and frequently over the most unfavourable terrain imaginable – ground which had been reduced to a swamp by rain and shells. Seventy-six tanks were available at Wytschaete,[9] and 216 at Third Ypres, but they achieved little, and the blame lies with the defective tactics which were forced upon them.

Did the other arms reap any greater rewards, after all their costly efforts? Nearly four weeks of drum fire, which consumed 93,000 tons of artillery ammunition, together with four months of heavy fighting and 400,000 casualties were the price of conquering a stretch of ground which measured at the most nine kilometres deep by fourteen kilometres wide. The Germans themselves lost 200,000 men, but they were able to prevent a breakthrough, and the U-boat bases were left undisturbed. The great sacrifice was squandered mindlessly, and yet it never occurred to the British high command that they were going about the offensive in totally the wrong way, and that it was quite impossible to conceal preparations for a push on this scale – giving the enemy time to take countermeasures, and presenting the British, after they had won every step of ground in such a costly way, with the prospect of having to overcome further lines to the rear. The command never appreciated that such a manner of fighting was no way to bring the war to a rapid conclusion.

Any notion of surprise or speed had to give way before the tenacious attachment to brute force and an unflinching application of fundamentally flawed methods. The terrain, the weather, the physical and moral resources of the army – and ultimately also of the British nation itself – became of little account in this unrelenting struggle. The blinkered mentality of the high command also accounts for its lack of vision in other respects – no change of tack at any price! For heaven's sake, no novel weapons![10] This was a period when the British Tank Corps, like the French *chars d'assaût*, stood within measurable distance of disbandment, for they could do no better than the infantry in these muddy battles in Flanders.

4. Mass Production

Although the British high command was averse to any innovation in tactics, the British and French armaments factories were set to work producing long runs of tanks, and the experience of combat in 1917 was evaluated in both the technical and tactical dimensions. By the summer of 1918 the British were supposed to have 1,000 tanks available for service, and the French 3,500. In addition the Americans intended to arrive in the theatre with 1,200 tanks, arranged in twenty-five battalions.[11] These figures were never actually reached because of difficulties with production, but the tank forces were now so large that they had the means of expanding success beyond narrow tactical limits into something of great operational significance. On the technical side the new models represented considerable advances in respect of cross-country mobility, range, speed, armour, armament and steering; tighter control was made possible by grouping tanks in battalions and companies.

The following data will give some idea of the tactical and operational capabilities of tanks in 1918:

Type	Mark V	Whippet	Renault
Weight (tons)	31	14	6.7
Armament	2 57mm	3 MG	1 37mm, 1 MG
Max speed (kph)	7.5	12.5	8
Reported range (km)	72	100	60

Before the high command could get its ideas in order about the deployment of tanks in 1918, whether for defensive or offensive purposes, there supervened an event in the late autumn of 1917 which presented the importance of tanks in an altogether new light. Even after the passage of years it is an episode which rewards our attention.

Notes

1. Guderian's almost total reliance on Swinton does cause some weaknesses in his account of the genesis of the tank. Swinton was one of the first to advocate an armoured, tracked, cross-country vehicle as an aid to the offensive on the Western Front. But Guderian does not make it sufficiently clear that the actual development of the tank in Britain had relatively little to do with Swinton. The War Office was over-worked and Lord Kitchener, the Secretary of State for War, had virtually no interest in Swinton's concept. After some rather half-hearted trials the War Office abandoned the idea. The early development of tanks, or 'Landships' as they were at first called, was carried out under the aegis of an Admiralty committee set up at the insistence of Winston Churchill, the First Lord. Swinton knew nothing about this committee until its work was quite advanced. For the Navy's side of the story see Rear-Admiral Sir Murray Sueter, *The Evolution of the Tank* (Hutchinson), 1937, passim.

2. Guderian's account is somewhat garbled here. Churchill was indeed an early enthusiast for some sort of armoured cross-country vehicle as a solution to the problem of attacking trenches and was aware of the existence of caterpillar tracks. He lacked detailed engineering knowledge, however, and did not specifically recommend the Holt system. As First Lord of the Admiralty, however, he was in a position to put to work on the problem men whose technical knowledge greatly exceeded his own. The successful rhomboidal landship design of Walter Wilson, employing tracks designed by William Tritton, was the result of this. A useful and concise account of the genesis of the tank in Britain can be found in David Fletcher, *Landships* (HMSO), 1984.

3. This document, 'Notes on the Employment of Tanks', February 1916, is quoted in full in Swinton's book, *Eyewitness* (Hodder and Stoughton), 1932, pp. 198214, and is available in the Stern Papers in the Liddell Hart Centre for Military Archives (LCMH) King's College London.

4. No charge of unwillingness to make use of new technology can be made to stick against Haig over the issue of tanks in 1916. Guderian's charge of impatience to use them derives from Swinton and is much more valid. For those interested in exploring further Haig's attitude to tanks in 1916 the relevant documents are: Haig to War Office, 9 February 1916 and 1 May 1916, WO 32/5754 and Haig to War Office 2 October 1916, WO 158/836, Public Record Office (PRO), Kew.

5. The order for 1,000 tanks is contained in Haig to War Office 2 October 1916, WO 158/836 (PRO).

6. Guderian is referring to General F. G. Anley. See J. F. C. Fuller, *Memoirs of an Unconventional Soldier* (Nicholson and Watson), 1936, p. 112.

7. Guderian's account of the early development of tanks in France seems to be derived largely from the French Official History and Dutil's *Les Chars d'Assaut*. There is still no scholarly account in English of the French development and use of tanks in the First World War.

8. The standard account of the initial British use of tanks on the Somme is Basil Liddell Hart, *The Tanks*, Vol. I (Cassell) 1959, pp. 71–81. Liddell Hart points out that the message printed in an English newspaper as having been reported by an air observer (and recorded by Guderian) that a tank was walking up the High Street of Flers with the British Army cheering behind it had a basis in reality but was rather embellished.

9. The battle which Guderian here refers to as Wytschaete will be more familiar to British readers as Messines. This limited action was one of the best planned and executed of all British attacks in the First World War. The contemporary official analysis of the performance of tanks at Messines is in the Kew branch of the Public Record Office (PRO) WO 158/858.

10. This very extreme denunciation of the British GHQ seems to be derived largely from Fuller's, *Memoirs*, especially pp. 13642 but there are also passages very hostile to GHQ in Swinton's *Eyewitness*.

11. On Allied plans in late 1917 to mass-produce tanks for the 1918 campaign the relevant British War Office file is (PRO) WO 158/813. Guderian omits to mention that by this stage in the war there was a considerable degree of inter-Allied planning in tank production matters. The most remarkable feature of this co-operation was a large, jointly financed Anglo-American tank factory which began to be established early in 1918 at Châteauroux in France. See Memorandum by Stern, 26 April 1918, Fuller Papers, Tank Museum, Bovington.

THE BIRTH OF A NEW WEAPON

1. CAMBRAI

Since their first action, in September 1916, the British tank forces had not only augmented their numbers, but undergone changes in organization and personnel. The original six companies developed into nine battalions, which from July 1917 bore the name of the 'Tank Corps'. In turn three brigades were formed of three battalions apiece. Each battalion comprised three companies of four platoons of four tanks each, and was supported by a mobile workshop.

The standard tank in the autumn of 1917 was the Mark IV. It resembled the Mark I of autumn 1916 in external appearance, but its armour was proof against the SmK round, and it carried an unditching beam which could be attached to the tracks to enable the machine to haul itself out of trenches. The tank weighed twenty-eight tons, and the 105hp Daimler engine gave it an average speed of three kilometres an hour, and a maximum of six. The crew consisted of an officer and seven men, and it was armed with two 58mm guns and four machine-guns in the 'male' version, or six machine-guns in the 'female'. It had a range of twenty-four kilometres. In November 1917 the British had ready for action 378 such Mark IVs, and 98 older models which were used as supply tanks.[1]

The Tank Corps was commanded by Brigadier (later Major-General) Hugh Elles, and his staff included Major J. F. C. Fuller as chief of staff, Major G. le Q Martel, and Major F. E. Hotblack, who was in charge of signals.[2]

After the Third Ypres offensive had so clearly failed, the leaders of the Tank Corps asked the high command for freedom to use the tanks in a more effective way. They were thinking along the same lines as Swinton's paper of February 1916, which had already been approved by the high command, but then forgotten in the next year.[3]

Three essential preconditions were laid down for the succcess of an attack by tanks: suitable ground, employment *en masse* and surprise. These deserve further exploration before we can proceed with our story.

Armoured forces are often criticized for the fact that they cannot be employed in every conceivable kind of terrain – most obviously they might find it impossible to cross high mountains, steep slopes, deep swamps or deep watercourses. But the same has been true of every other vehicle in history – it is just a question of using whatever means lie at hand, for the lack of anything better, and if necessary we will have to tackle obstacles of this kind by making artificial passages or simply flying over them. It is true that technology is striving constantly to improve the cross-country mobility of military vehicles, and tanks in particular; a great deal has been achieved very recently, and we have every confidence that still more significant advances will follow. However terrain remains a consideration which will always have to be borne in mind.

It is no good sending armour into an attack over ground where it can have no real hope of making progress. It is every bit as wrong to lay down a preparatory bombardment which reduces the terrain to a lunar landscape, where even the most effective modern machines – not to mention horse-drawn vehicles – will end up getting stuck. If the tanks are to keep on the move they must be spared the problem of having to overcome broken terrain when they are on the attack. We are very fond of designating 'axes of advance' [*Gefechtsstreifen*], but they must not be drawn with geometrical rigidity across hill and dale, across rivers and woods; for the tanks, at least, we must take account of the conformation of the ground and the nature the of the surface. If the good going for the tanks does not happen to suit the infantry and artillery, it may be necessary to direct the attack of the armour along an axis which runs obliquely to that of the infantry. The main thing is for the tanks to get at the enemy.

The question of suitable terrain is linked intimately with deployment *en masse*. As we have seen from our historical examples, decisive success has never been achieved by sending in tanks as small units, for whatever the reasons – perhaps there were not many machines available in the first place, or perhaps the command decided to feed in a large number of tanks in penny packets, as happened with the French on 16 April 1917. The outcome was the same – the enemy always had time to organize an effective resistance. Tanks in the World War were slow moving, and you could finish off a tank attack just by laying down a concentrated fire of artillery.

The effect of artillery is much more patchy when it has to cope with a large number of tanks attacking simultaneously, and this is true whether we are talking about the artillery of the World War or the present-day anti-tank guns. But if we are to commit tanks *en masse* we come back to our first point – we must in turn have suitable terrain for our attack.

Surprise is the third precondition for a thoroughgoing success on the offensive. Since time immemorial there have been lively, self-confident commanders who have exploited the principle of surprise – the means whereby inferior forces may snatch victory, and turn downright impossible conditions to their own advantage. The effects on the morale of both parties are immense – but this very element of incalculability proves a deterrent to ponderous spirits, and this is probably why they are so reluctant to embrace new weapons, even when the inadequacy of the old ones is all too clear.

Surprise may hang upon the very novelty of the weapon in question. It takes considerable daring on the part of a commander to use a weapon for the first time, but in the event of success the rewards will be all the greater. We have seen, however, that neither the Germans with their poison gas nor the British with their tanks were willing to accept the risk of employing new weapons *en masse* in a surprise attack. Once the fleeting opportunity had been missed, surprise could depend only on the kinds of time-honoured techniques which had been used with the conventional weapons. Even then, however, there remained considerable scope for catching the enemy off guard.

Apparently superficial technical advances may trigger surprises which can prove extremely painful for the enemy. Twenty years had passed since the Prussians had gone over from their muzzle-loading infantry weapons to their breech-loading needle gun, and yet it proved the key to victory in 1866 – their Austrian enemies had not appreciated the magnitude of the technical advance, and they were astonished at the lethal effect of the needle gun on the battlefield. Then again the German 42mm mortar represented only an increase in calibre on a type of piece that was already well established, but in 1914 it smashed through the armour and concrete cover of the Belgian fortresses which had been reputed to be impregnable. And yet the 42mm mortar, like the needle gun, had been tested only on peacetime ranges before it was taken on campaign. There was no question of waiting to see whether other armies were trying similar weapons in their own wars, or indeed of establishing whether the foreigners had any such weapons at all – that would have sacrificed surprise. On the contrary, the 42mm mortar was a carefully guarded secret, and in the event surprise was total.

The same holds true of tanks in the last war. The prospects for surprise did not vanish altogether even a year after the machines had first appeared in the field. After all, there was no guarantee that the Germans would actually exploit the opportunity to come to terms with a possible enemy use of tanks *en masse*, or likely improvements in design and tactics. The potential for surprise had diminished only a certain amount, and the exact degree depended very much on the Germans.

After the conventional weapons had failed, and one year after tanks had been misused despite the protests of Swinton, their creator, the British high command at last yielded to the tank officers, and put at their disposal those battleworthy units that remained from the forces that had been squandered at Third Ypres. For the first tank battle in history General Byng's Third Army was assigned: two corps of two divisions each; one cavalry corps of five cavalry divisions; one Tank Corps of three brigades of two battalions each; 1,000 guns and a large number of aircraft.

And that was all. It fell short of what was needed to accomplish a major breakthrough, even if the British managed to achieve surprise, and the only divisions that the Germans could put against them were burnt-out ones. Ultimately, the British did not have the necessary reserves. As for the scope of the attack, the plan envisaged a breakthrough between Gonnelieu and Havrincourt on a frontage of two corps with the help of tanks, thus opening a way for the cavalry to pass through and exploit the success. The British evidently intended to take Cambrai, but it is uncertain whether they had any further objectives.

The terrain in question stretched to the north-east between Gonnelieu and Havrincourt, and suited the offensive well enough. The rolling and predominantly open ground descended gently to the River Schelde, which secured the right flank of the attack, as it flowed from Banteux to Crèvecoeur, passing to the east of Gonnelieu; the river then described a sharp bend as it veered from north-east to north-west across the area of the attack by way of Masnières, Marcoing and Noyelles, and finally followed a gentle curve as it inclined north-eastwards again in the direction of Cambrai. The Schelde and the parallel Scarpe Canal were passable only by bridges. In front of the left wing the villages of Fontaine-Nôtre-Dame and Bourlon, together with the intervening Bourlon Wood, formed a kind of bastion which posed potential difficulties for the tanks. Between this bastion and the Schelde the only obstacles in the way of the attack were presented by the various villages. However the walls and cellars offered the defenders good protection against tanks, and demanded special attention on the part of the British if they were to be captured or suppressed.

The British knew that the sector chosen for the attack was held essentially by 54th Jäger Division and that, quite apart from the Tank Corps, the British had a six-fold superiority in infantry and artillery alone. The British III Corps with its 12th, 20th and 6th Divisions was going to attack to the east of a boundary stretching from the western edge of Ribécourt-la-Tour to the western edge of the Bois des Neufs; the 51st and 62nd Divisions of IV Corps were to attack to the west of the boundary. The first objective was a phase line extending from la Vacquerie to north of Havrincourt, passing by way of the railway to the north of Ribécourt; the second phase line ran from le Papé to north of Flesquières, and the third from la Justice to Graincourt, passing to the south-west of Cantaing. As it advanced towards the Schelde III Corps would take over the protection of the northern flank, while IV Corps continued its push in the direction of Fontaine-Nôtre-Dame. The 56th Division was to carry out a feint attack against the defences adjoining to the left between Quéant and Inchy so as to divert the attention of the Germans. Further diversionary attacks were to be launched to the right of the genuine attack, at the farm of Gillemont and to the left at Bullecourt. The 29th Division was ordered to follow up behind III Corps' attack as reserve, and seize the line Masnières–Rumilly–Marcoing.

Finally the cavalry would exploit the success. The 2nd and 3rd Cavalry Divisions were to execute a flanking attack to the south and east of Cambrai, and 1st Cavalry Division was to do the same but to the west. In the process 1st Cavalry Division was to help the infantry to capture Cantaing and Fontaine-Notre-Dame (respectively north-west and north of Cambrai), cut off Cambrai itself and unite with the cavalry working around to the east of the town. There were plans to push units farther north to the Sensée stream, so as to disrupt the German rearward communications.

This time the artillery departed from convention. The attack was to be ushered in by a single artillery strike, in place of the lengthy preparatory fire and meticulous registration. The German batteries, command posts and observation posts would be suppressed or blinded with smoke, while long-range guns shot up the approach routes, villages and railway stations in the German rear areas. In addition a creeping barrage would fall in front of the attacking forces. The artillery reached its battery positions undetected by the Germans.

There were about as many aircraft as tanks, and the pilots and observers were told to keep a sharp eye out for enemy reserves, and report any impending counter-attacks without delay. The tanks were fully integrated in the plan of attack and the units were assigned to the various formations as follows:

III Corps
12th Division – two battalions, with 48 tanks in the first line, 24 in the second and twelve in reserve;
20th Division – two battalions minus one company, with 30 tanks in the first line, 30 in the second and eighteen in reserve;
6th Division – two battalions, with 48 tanks in the first line, 24 in the second and 23 in reserve;
29th Division (following up as reserve) – a single company, with twelve tanks in the third line and two in reserve.

IV Corps
51st Division – two battalions, with 42 tanks in the first line and 28 in the second;
62nd Division – one battalion, with 42 tanks in the first line and fourteen in the second.

Each unit was assigned a specific task, with the platoon counting as the smallest sub-unit. Some of the units were told to break off as soon as they could and tackle the greatest threat of all – the German artillery, which was also going to be taken under attack by bomber aircraft.

Some of the features of the attack were rehearsed with the infantry beforehand. Fascines were prepared and brought up on tanks, for crossing the broad German trenches. In detail, 'male' tanks armed with guns would advance up to the defences, crushing the obstructions and suppressing the troops by fire, whereupon a 'female' would throw its fascine into the trench. The little convoy would clamber across the line at this point and repeat the manoeuvre at the next trench. Captured trenches were to be kept clear by fire until the British infantry arrived to take them over.

Even the British troops were kept in the dark about the purpose of the preparations. The Tank Corps was assembled at Albert on the pretext of training, and two nights before the attack moved forward to assembly areas behind the front, mainly in the Bois d'Havrincourt. On the last night it advanced to the start-line immediately behind the foremost trenches. The gloomy November weather hindered German reconnaissance.

Since March 1917 the Germans had been standing in the Hindenburg Line. This defensive line had not arisen by chance out of previous fighting, like the defences on the other sectors, but had been constructed after meticulous surveying and the experience of two years of positional warfare. Nearest to the enemy extended an outpost trench protected by belts of barbed wire. Next a number of strongpoints were established in the space between the outpost line and the first battle trench proper, which was more than three metres broad and furnished with a large number of dugouts. Like the second battle trench three hundred metres to the rear, it was covered by barbed wire obstacles averaging thirty metres deep. Both battle trenches had good fields of fire, and a proliferation of communication trenches enabled the Germans to move inside their defensive system under cover. Some two kilometres behind this first position, a second position had been laid out but not fully completed because of the lack of labour. It comprised merely a series of hasty works extending from the 'enemy' side of Bourlon Wood to the Bois des Neufs, and from there to the north bank of the Schelde. It is worth noting that Cambrai was considered a quiet sector where battle-weary divisions from Flanders could recover.

In November 1917 the German forces on this tranquil stretch of the front consisted of 'Army Group Caudry' which stood under the command of XIII Corps. The 20th Landwehr Division was deployed on both sides of the Cambrai–Bapaume road; the 54th Jäger Division occupied a frontage of eight kilometres between Havrincourt and la Vacquérie; the 9th Reserve Division adjoined to the south. The three infantry regiments of 54th Jäger Division were arrayed shoulder-to-shoulder with two battalions each in the front line, and a third occasionally resting. Only the first position was completed, the intermediate position remaining empty.

As late as 16 November 1917 the commanders of Second Army had the impression that no major attacks were to be anticipated in the immediate future. Reconnaissance patrols on 18 November confirmed that the British 36th Division was holding the front at Trescault, as before. The prisoners taken on this occasion said that the division was due to be relieved by 51st Division, and that they had caught sight of tanks in Bois d'Havrincourt; they added that an artillery preparation of several hours would precede an attack that was planned for 20 November. On the 19th a prisoner confirmed that the British 20th Division was still present. Enemy air activity and ground traffic were livelier than usual, and a number of new batteries were espied in the Bois d'Havrincourt. Otherwise 19 November passed peacefully, and there was no notable increase of registration fire on the part of the British batteries.

The Germans took a number of countermeasures on the basis of the available reports, even though there was no indication of a major attack in the old style, and in fact prisoners taken elsewhere repeated the same stories of intended attacks on their sectors as well. Late on the evening of 19 November the Germans ordered a higher state of alert, and 54th Jäger Division's artillery laid down targeted and harassing fire on the nearest

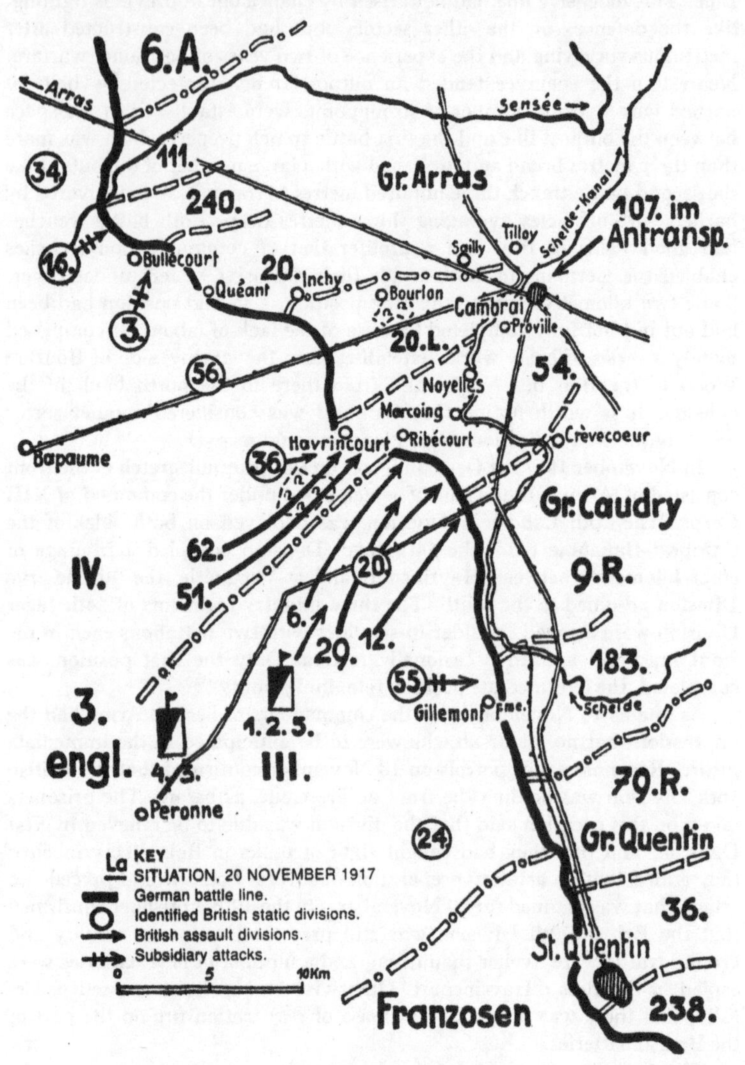
TANK BATTLE AT CAMBRAI, NOV. - DEC. 1917
6.A.
Arras
Sensée
111.
34
240.
Gr. Arras
Scheide-Kanal
107. im Antransp.
Tilloy
Sailly
16.
Bullecourt
20.
Inchy
Quéant
Bourlon
Cambrai
3.
Proville
20.L.
56
54.
Noyelles
Marcoing
Crèvecoeur
Bapaume
Havrincourt
Ribécourt
36
Gr. Caudry
62.
IV.
20
9.R.
51.
6.
29.
12.
183.
55
Scheide
Gillemont
Fme.
3.
engl.
4./5.
1.2.5.
III.
79.R.
Peronne
24
Gr. Quentin
KEY
SITUATION, 20 NOVEMBER 1917
German battle line.
Identified British static divisions.
British assault divisions.
Subsidiary attacks.
0
5
10Km
36.
St. Quentin
Franzosen
238.

Sketchmap 8.

enemy trenches, and artillery strikes on Bois d' Havrincourt, the village of Trescault and the approach routes. The high command placed the left-hand regiment of 20th Landwehr Division, which was holding the Havrincourt sector, under 54th Jäger Division so as to ensure unity of command on the likely scene of action. Likewise Army Group Caudry took under command 27th Reserve Jäger Regiment, the headquarters of an artillery detachment, and two batteries – all of which were brought up from army reserve. It was given out that reinforcements of batteries of heavy artillery were to be expected for the 20th. The 27th Reserve Jäger Regiment was detailed for counter-attacks, and placed behind the two right-hand regimental sectors of 54th Jäger Division: part of its first battalion was assigned to 84th Jäger Regiment and pushed forward to Flesquières, and part was lodged in Fontaine-Nôtre-Dame; the regimental headquarters and the second battalion were positioned in Marcoing; the third battalion remained in Cambrai as group reserve. In addition 54th Jäger Division received two field artillery detachments from 107th Jäger Division which had just arrived from the Eastern Front; these were placed at Graincourt and Flesquières.

The Germans on the whole do not seem to have reckoned on a major British push; they had every confidence in the strength of the Hindenburg Line. All the more remarkable were the speedy and energetic measures of defence which, as we have just seen, were undertaken on the initiative of Second Army, Army Group Caudry and 54th Jäger Division. It was unfortunate, however, that they neglected specific precautions against an armoured attack: there were no guns in position that had an arc of fire capable of tackling tanks at short range, and it seems that the infantry learned only very late about the possibility of a tank attack, and consequently had very little SmK ammunition when the assault arrived.

Dawn on 20 November came with a greyish light. There was a false alarm at 0600; a barrage descended at Havrincourt and then everything fell quiet again. At 0715 the British artillery barrage pounded the German positions, and all our troops took cover in the dugouts, leaving only sentries outside. On past experience several hours would now elapse before the enemy infantry attacked, and the German artillery fired no more than a feeble barrage ahead of our outpost line into the smoke and mist of that pale morning. The outposts were taken by surprise when suddenly indistinct black forms could be discerned. They were spitting fire, and under their weight the strong and deep obstacle belt was cracking like matchwood. The alarm was transmitted to the men in the trenches, and the troops hastened to their machine-guns and tried to put up a defence. It was all in vain! The tanks appeared not one at a time but in whole lines kilometres in length! The SmK ammunition proved to be ineffective, the barrage could not be brought back from its impact zone in front of the outposts, and there were few grenades available – and fewer still which did any damage to the enemy machines as they kept up their fire. Effectively the German infantrymen were defenceless, pinned down and unable to withstand the mighty material superiority of the British. The only alternatives were death or surrender, since nobody could make off to the rear and hope to survive under this fire.

Surely the reserves will counter-attack and help us out! The 54th Jäger Division duly ordered the commander of 27th Reserve Jäger Regiment to counter-attack with two battalions and recover the lost trenches; the mission was exactly as laid down in regulations. But the headquarters of 108 Infantry Brigade was in charge of the infantry battle, and it immediately countermanded the move; the counter-attack fell through, and the only unit that could lend any help was 3rd Battalion, 27th Reserve Jäger Regiment which had been standing in group reserve at Cambrai.

A number of other units were brought up from 107th Jäger Division, recently arrived from the Eastern Front. At 0940 two battalions were set in motion by way of Masnières, and a further battalion in the direction of Crèvecoeur, and they were placed under the respective commands of 54th Jäger Division and 9th Reserve Division; another regiment was pushed to Fontaine, Cantaing and Proville and put at the disposal of the Army Group Caudry; the third regiment was placed in Cambrai as army reserve.

The reports from the front line remained scanty in the extreme. The ground mist ruled out aerial reconnaissance, and the British creeping barrage rolled forward and impeded observation.

Meanwhile unit commands did their best to follow orders and put up an aggressive defence. The 2nd Battalion, 27th Reserve Jäger Regiment, advanced from Flesquières towards Havrincourt along two communication trenches. There were no accurate reports as to what was going on, though the wounded talked of a great number of tanks. Elements of the companies then left the trenches in order to deploy across open ground. In its eagerness to get forward the battalion ran into an attack by tanks and was largely wiped out. The 387th Landwehr Jäger Regiment, to the left of 84th Jägers, was overrun and broken through; the adjoining 90th Reserve Jäger Regiment fared no better, and even its headquarters fell into the hands of the enemy. A slight relief came only when visibility improved and the tanks came within range of the artillery positioned around Marcoing.

The 19th Reserve Jäger Regiment, forming the right wing of 9th Reserve Division, was likewise hit by the tank attack and suffered very heavy casualties, though it continued to hold Banteux and the line of the canal.

Within a short time the whole defensive system had been lost along the entire frontage of the British offensive. Spearheaded by the tanks, the attack now flooded over the intermediate position. There was just one exception, the village of Flesquières, where the Germans managed to hold firm, thanks to the solid construction of the houses and the shelter available in the cellars which afforded a measure of protection against armour. Moreover the commander of 27th Reserve Jäger Regiment, Major Krebs, had taken over the command at 0900 and given a number of apposite orders – for a start he called a halt to those senseless and bloody counter-attacks against armour by unprotected infantry. Krebs was now able to hold back at least the machine-gun company and a rifle company of his 2nd Battalion, together with half of 1st Battalion, which was supposed to have attacked regardless.

He established these forces in and near Flesquières, together with the other half of 1st Battalion which had come up on trucks from Fontaine-Nôtre-Dame and debussed south-west of Cantaing. Elements of 84th Jäger Regiment and 108th Pioneer Company came under his command, and the Germans tied together hand-grenades in bundles as charges. It was owing to this clear-headed direction, to the devotion of the six hundred defenders, and above all to the magnificent support from the batteries of 108th and 282nd Field Artillery Regiments that Flesquières was held until nightfall.

We may note here that the defence of Flesquières shows that infantry are perfectly capable of holding a great variety of locations against armoured attack, provided that those places are properly evaluated and exploited; conversely unsupported armour cannot always be guaranteed to wipe out defending infantry.[4] We shall have more to say on this point later on.

At 1050 the commander of Army Group Caudry once more freed its reserve – 52nd Reserve Jäger Regiment – and ordered 54th and 107th Jäger Divisions to desist from further counter-attacks, and instead hold their existing positions to the bitter end. Trucks were made available to bring up the infantry, and cars for the headquarters. No reinforcements could be expected before evening, however, and the situation remained critical in the extreme. As General Ludendorff commented, 'We suffered an acutely painful lack of trucks with which to bring up our troops.' (Ludendorff, *Meine Kriegserinnerungen*, 393, 395.) The recruits from the training depots of the divisions in question were rushed up at breakneck speed, and thirty men from 54th Jäger Division's HQ were actually sent to hold the Schelde Canal.

In this crisis the commander of 18 Reserve Infantry Brigade (of 9th Reserve Division), Colonel von Gleich, was cool-headed enough to hang on to the existing bridgeheads over the Schelde Canal, for he knew that they could be useful for a German counter-attack. The villages of Banteux and Honnecourt therefore remained in German hands, and northwards from Banteux to Crèvecoeur 9th Reserve Division was able to hold a shallow front along the western bank of the canal.

Marcoing fell to the British, after a brave defence on the part of a number of batteries, and the enemy tanks were able to cross the canal at this point. The German survivors fell back on Cantaing, though they had no hope of being able to hold it against a serious attack. Noyelles was lost, although the Germans were able to blow up the bridge, which was not the case elsewhere. The remains of 3rd Battalion, 27th Reserve Jäger Regiment, occupied the eastern bank of the canal between the sugar factory and the Ferme du Flot; farther south 227th Reserve Jäger Regiment of 107th Jäger Division was able to stop the enemy when they were in the process of crossing the canal at Masnières. A Canadian cavalry squadron, spurring towards Cambrai, attacked a German battery, but was in turn thrown back with heavy losses by the troops of the field recruit depot of 54th Jäger Division who arrived on the scene at that juncture. The Germans had precious few reserves available, however, and there was little hope of holding this sector either, once the British made up their minds to attack in force. Incomprehensibly this attack never materialized.

Not until late in the afternoon did further reinforcements arrive from 10th Jäger Division in Cantaing, but they came just in time to break an attempt on the part of British 1st Cavalry Division to push to the north on horseback. The Germans now formed a new line of defence at Anneux and Cantaing, and at 0415 on 21 November Major Krebs was able to carry out an undisturbed evacuation of Flesquières, which had been defended so bravely.

For the Germans the night of 20/21 November was a time of disquiet and uncertainty. Would the enemy recognize and seize their opportunity? Did they have the reserves to carry out a breakthrough on the operational scale? The German command strove to find out where the front line actually ran, and to restore some kind of order amid the confusion of its units.

Having described the course of events from the German side, we turn to the viewpoint of the attackers.

We have shown how the tanks were distributed among the assaulting divisions. Essentially the tanks would be attacking in two lines, with only one of the companies forming a third line, being attached to 29th Division which was following up as a reserve. The two lines of tanks were tied to the corresponding lines of infantry, with the result that the second line of tanks was not at hand to help out the first in the event of trouble, or exploit any opportunities which the first line might create. Altogether the disposition of the attack was strictly linear, devoid of depth and devoid of tank reserves. Once the resources of the Tank Corps were expended its command was virtually neutralized – like the direction of the army as a whole. General Elles could do no more than direct whatever tanks he could from his place in the leading tank of the central brigade.

At 0710 the tank assault left its start-line, one thousand metres from the enemy. At 0720, at the same time as the tanks were crossing the front line, the British artillery strike and creeping barrage descended in a mixture of high-explosive and smoke shells. The smoke blinded the German artillery, but it also occasioned some disruption to the order of the tanks, which were often reduced to finding their way by compass. The Hindenburg Line had been accounted the strongest field position on the Western Front, but now the British took it with effortless ease. They smashed through obstacles, crossed the trenches with the help of fascines, and destroyed or captured all the troops holding the forward position – and the same fate was shared by the German reserves when they arrived with their counter-attacks. The British followed in a systematic way behind the slow march of their creeping barrage, but by 1100 they had succeeded in reducing the whole of the German defensive system with the exception of Flesquières. The gun duel ended with the defeat of the German batteries, but the pace of the attack remained slow, and the tanks suffered severely whenever the German battery commanders were active enough to haul a few guns out of their dug-in positions and fire over open sights.

Towards noon a decisive victory was within the grasp of the British. A considerable stretch of the Schelde Canal was in their hands, from Crèvecoeur to south-west of la Folie, together with most of the bridges. Wide gaps had opened in the German front between Crèvecoeur and Masnières, and again between Cantaing and Flesquières. To the east of Moeuvres the attacks had begun to roll up 20th Landwehr Division to the north. Altogether, apart from a few pockets of resistance, the German defences had ceased to exist on a frontage of twelve kilometres; the breakthrough had succeeded, and everything now depended on whether the attackers could exploit the opportunity. Every hesitation, every delay offered the defenders the possibility of bringing up reserves, forming a new front line and putting the victory once more in doubt, after it seemed to have been won so easily and speedily. As regards fresh infantry, the British had at their disposal only 29th Division; the only uncommitted tanks were the twelve machines attached to that division. However the British high command also had available an entire cavalry corps of five divisions, which was prized as a weapon particularly suited for exploiting a success. Broad lanes for the benefit of the cavalry had already been forced through the German obstacles; thirty-two supply tanks had been entrusted with this task, and two more had brought up special bridges for the cavalry. From 1330 thirty combat tanks had been waiting impatiently for the cavalry at Masnières, and 29th Division was doing the same in the bridgehead at Marcoing. It was just a question of following through the drama to the final act. In the event the cavalry did not appear until about 1630 in the form of a single squadron at Masnières, and somewhat stronger forces at Cantaing, and they experienced a bloody repulse at the hands of the weak units of 54th and 107th Jäger Divisions. In fact the British had been attempting to get cavalry to act in concert with tanks – to employ it at long last in the mobile mounted role on the Western Front – and the attempt had failed. The chance of victory had come and gone; five cavalry divisions had been unable to rip through a thin screen of a few machine-guns and rifles.

By the evening of 20 November the first tank battle in history had come to an end. In a few hours the strongest position on the Western Front had been broken on a frontage of sixteen kilometres and to a depth of nine kilometres. Eight thousand prisoners and 100 guns had been captured, for a British loss of 4,000 men and 49 tanks. Cambrai was a great British victory, and the bells rang out in London for the first time in the war. The tanks had achieved a breathtaking successs, and fully justified their existence. Swinton and Elles exchanged telegrams of congratulation.

But what did the British then do with the breach they had made in the German front? There was every reason to expect a second and even greater blow, and a number of measures were actually taken – two divisions were diverted to General Byng from forces destined for Italy, while the French forwarded a reserve group of two infantry and two cavalry divisions under General Degoutte to the neighbourhood of Péronne by rail and truck. The British reserves were fed into action by driblets, however, and the French

TANK BATTLE AT CAMBRAI, NOV. - DEC. 1917

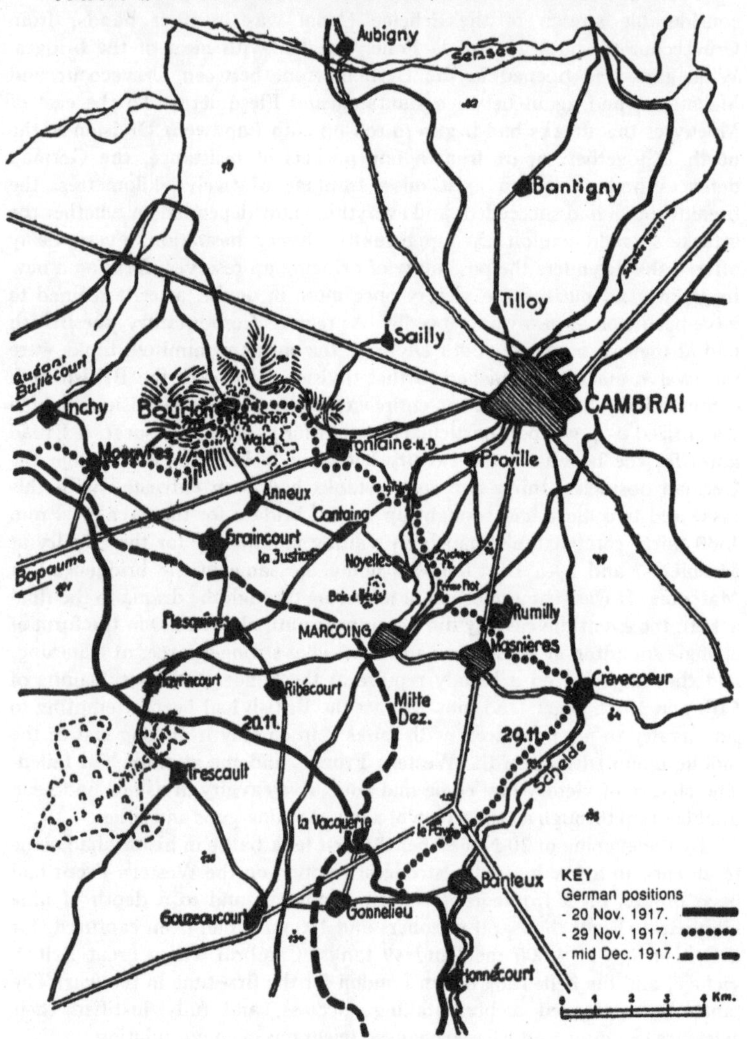

Sketchmap 9.

did not materialize at all. Even the Tank Corps could field only part of its strength for the fighting which lay ahead.

Meanwhile German reinforcements were pouring in from all sides. The situation on 21 November was still critical in the extreme, and the commander of Army Group Caudry reported in the morning: 'I cannot conceal the seriousness of the position, if the enemy keep up their attack with tanks before we receive more reinforcements in artillery. In that case we will have no means of withstanding a further break-in and ultimately a full breakthrough.' The 20th Landwehr Division had lost about two-thirds of its establishment, and 54th Jäger Division was virtually wiped out. Once more the Goddess of Victory beckoned the British on. And yet, if only Third Ypres had never happened! If only it were possible to bring back to life all the divisions which had been sacrificed in that stupendous grinding battle! Twenty-two million pounds had been expended on ammunition in that struggle – if only they had been devoted to tanks instead!

The British spent the morning of 21 November preparing to launch the new attack. It was badly co-ordinated, and when it was delivered it was supported by only forty-nine tanks and achieved only a limited success. And yet the opportunities were there. In the afternoon a number of British battalions advanced betwen Graincourt and Marcoing, actually marching in close order to the sound of music and with their officers on horseback, without being shot up by the German artillery. A number of tanks took Cantaing and penetrated Fontaine-Nôtre-Dame, but the infantry failed to exploit, even though a gap yawned between the latter village and Bourlon Wood during the following night. A few machines tried to push from the railway station at Marcoing against Cambrai, but a German battery arrived in the nick of time and drove them off with some loss.

On 23 November and during the following days sixty-seven tanks were able to participate in the attack, which focused mainly on the Bourlon massif. By 27 November the Germans still held the villages of Bourlon and Fontaine-Nôtre-Dame, but the British now had Bourlon Wood itself. Although only a few tanks had been able to join in this last push, it had stretched the nerves of the defenders to breaking-point, and panic was averted only by the intervention of a number of particularly vigorous and determined commanders. The fact that these officers succeeded is nevertheless to the credit of the troops concerned.

On 27 November the British began to withdraw the tanks behind the front line for major refits, some of the units departing by train. This was the day when General Ludendorff announced at a conference at le Cateau that he had decided to launch an immediate counter-attack. The preparations went ahead so swiftly that the Germans were able to subject Bourlon Wood to a gas attack on the 29th, and assault it the next day after a one-hour bombardment. The British were taken by surprise, especially on their southern sector, and they not only had to throw in their cavalry but retrieve with some urgency the tanks that were already in the process of being transported away. On the night of 4/5 December the Germans recovered

Bourlon Wood, after a tough fight, and by 6 December they had not only regained large areas of the ground they had lost, but south of la Vacquerie they took ground well beyond their old front line. If the Germans were unable to push their success any farther, it was because of their lack of reserves, the diminished battleworthiness of their divisions in the course of the fighting, and the poor organization of their supply system. Altogether the Germans took 9,000 prisoners, 148 guns and the more than 100 tanks which had been abandoned on the battlefield more or less damaged since the battle opened on the 20th; the British on their side announced the capture of 10,500 prisoners and 142 guns. The German disgrace of 20 November had been brilliantly expunged.

Before we turn to the lessons of Cambrai it is worth pausing to consider the overall British losses from 20 to 30 November 1917:

III Corps: 672 officers, 5,160 men;
IV Corps: 686 officers, 13,655 men;
Cavalry: 37 officers, 674 men;
The Tank Corps on 20 November: 118 officers, 530 men;
2 Tank Brigade from 20 November to 1 December: 67 officers, 360 men.

These figures show how the intervention of the tanks at Cambrai gained the same extent of ground as the British had won at Ypres, but with far fewer losses and in an incomparably shorter time. The statistics also indicate how the nine weak battalions of the Tank Corps had fought with the greatest courage, and had shrunk from no sacrifice in their striving for victory.[5]

We turn now to the lessons which the attackers and defenders did learn, and also the ones which they ought to have learned.

The British concluded that as a weapon the tanks had fulfilled their missions magnificently. However, a number of technical improvements were demanded – one-man steering, a more powerful engine, and a greater obstacle-crossing capability, together with the need for a new type of fast tank for exploiting successful breakthroughs. In 1918 the first three requirements were met in the Mark V, and the last in the Medium Mark A ('Whippet') and the armoured car. On the organizational side the Tank Corps was built up to five brigades with thirteen battalions. New machines were delivered in the course of the winter, and the British were able to put in some useful training.

In expectation of a German offensive in the spring of 1918, the question arose as to how the Tank Corps could best be employed in the defence. There were two possible courses of action. One was to hold back the corps in army reserve until the direction of the main German thrust had become clear, and then employ the tanks as a united formation on the counter-attack; this solution was supported by the success the tanks had achieved when they had gone into action *en masse* at Cambrai, compared with the earlier failures when they had fought as small units. The alternative was to break up the Corps into small troops which would be stationed behind the front line as local reserves – at the risk of having a large number of tanks standing around inactive in the hour of need on the quiet sectors, while not enough machines were available to meet the enemy on the location of the main break-in. The British high command chose the second course, with the result that when the German offensive arrived the tanks scored only local successes; the command drew the erroneous conclusion that tanks were not really battleworthy. So-called experts declared that it was impossible to repeat the 'once-in-a-lifetime' surprise victory at Cambrai, and cited as evidence the supposed failure of armour against the German spring offensive. They neglected to say that they themselves were responsible for the failure in question. The intended augmentation of the Tank Corps was deferred, and some of the units were actually disbanded to make up the losses in infantry.[6] A reassessment in favour of the tanks took place only after the action at Hamel on 4 July 1918.

As we have seen, the British overlooked a number of potentially valuable lessons from their initial victory at Cambrai. For our purposes, however, it is worth stating that:

– the extraordinary success of the tanks at Cambrai came from the fact that for the first time ever they had been employed *en masse* on a broad front;

– the success would have been greater still if the tank attack had been endowed with greater depth, if mobile and effective reserves had been available and if, instead of being content with taking the forward German positions, the British had striven from the outset to strike to the full depth of the defensive system, eliminating batteries, reserves and headquarters in one blow, and if, finally, the air force had been brought in to lend extensive tactical support.

The success of the tanks fell off markedly as soon as they were forced to attack individually and in small units – a particularly dangerous practice now that the enemy were beginning to come to terms with them. The tank losses piled up, and the fewer the tanks that were available the more the infantry came under enfilade fire from both flanks, and the less they were able to exploit whatever successes the armour had achieved. The high command had clung to some misconceptions about the worth of the conventional arms, particularly on the offensive, and it persisted in flinging them *en masse* into a series of bloody and abortive assaults; the new weapons, by way of contrast, were almost invariably committed one at a time, bit by bit, and scattered over a wide frontage. The high command then asked itself why so little had been achieved when so much had been expected.

Beyond any doubt the Germans had sustained a painful blow on 20 November. Their losses over the following days were also substantial, even when the tanks were attacking individually. The existing obstacles proved unavailing against tanks, and the German artillery tactics were useless. Now that the enemy were able to roll over German trenches without artillery preparation, they had nullified the tactic of the long-range defensive barrage. The figures of German missing indicate that the infantry were too often helpless; frequently it was a question of re-assembling the remains of the regiments as they streamed back from the front without their weapons.

It is clear that the tank was a battle-winning weapon when it was employed *en masse*, as at Cambrai, and now in 1918 the Germans had to reckon with the appearance of far greater numbers and improved designs. There were two things which could be done: bend every effort to augment the defensive capacity of the troops, and then create our own tank forces – especially if we intended to go over to the attack ourselves. We shall consider both courses in turn.

On the defensive side the Germans devised single-shot, anti-tank rifles and anti-tank machine-guns, firing 13mm bullets. In the event the machine-guns were not ready by 1918, and only the rifles appeared in the front line. Minenwerfers were fitted with carriages for low-trajectory fire, and each of the armies on the Western Front were assigned ten anti-tank guns which were mounted on commercial trucks. Tank traps and a number of anti-tank minefields were emplaced, and in general more emphasis was put on protection against tanks when new positions were being laid out. Artillery tactics too began to change, as witness the deployment of individual guns well forward for anti-tank defence, and the practice of attaching batteries to the regiments of assaulting infantry. Direct, aimed fire by single rounds was given more significance, as opposed to mass bombardments.

It must be admitted, however, that these measures did not amount to very much, and yet the Germans did still less about creating a tank force on their own account. The War Ministry gave Number One Priority to accelerating the production of the A7V tank, and repairing the British tanks captured at Cambrai, but that was about as far as it went. The armour available in 1918 consisted of no more than fifteen A7Vs and thirty captured British tanks, and they had few spares to back them up – no decisive blow was possible with a force of these modest dimensions. In fact the German infantry had performed so magnificently in the counter-attack at Cambrai on 30 November, that the German high command attached no great significance to the offensive potential of tanks. It distrusted them anyway, and after the difficulties the Germans had experienced with logistic support on that occasion it seemed more worthwhile to acquire supply trucks with cross-country capability. A number of the existing tank workshops were converted to this end, and a number of the so-called *Marienwagen* duly put in an appearance, though in insufficient quantity.

The success of the counter-attack on 30 November confirmed the German high command in the conviction that on the Western Front, just as in the East, infantry and artillery by themselves had the striking power to break through enemy positions, as long as the attack came as a surprise and was launched in sufficient breadth and depth. This belief was to a large extent justified; but just one consideration was overlooked in all the planning which went into the mighty spring offensive of 1918. If, as seemed likely, the conventional arms managed to break into the enemy zone of defence, would they have the speed to exploit and expand the break-in to a full breakthrough? In other words, did the available means permit a tactical success to be developed into an operationally significant victory? The question was all the more relevant because the enemy had the ability to rush up troops to fill the gaps. Would the Germans also be able to cope with the likelihood of even larger numbers of enemy tanks, capable of higher speeds and longer ranges than before?

The events of 1918 were to give an unequivocal answer to all these questions.

2. 1918. THE GERMAN SPRING OFFENSIVE. SOISSONS AND AMIENS

After meticulous preparations the German Army concentrated all its strength in one mighty effort to break the deadly embrace of the Allies by victory in the field on the Western Front. There was no other way out, after the period of unrestricted U-boat warfare had failed to produce the desired effect, and diplomatic efforts had proved fruitless. The supreme command was in no doubt as to the magnitude of its task. General Ludendorff had made this repeatedly clear in the preliminary conferences, and he addressed himself with unflagging activity and inexhaustible application to driving forward the preparations he believed were essential to the success of the offensive. As it embarked on this decisive battle the whole army reposed its complete trust in the high command. The troops identified themselves with the purposes of the leadership, and shared the resolve to master a task which, in sheer human terms, was impossible.

In line with the tactical thinking of the time, success was to be achieved by an abbreviated artillery preparation, and completed by the infantry, who would attack according to a number of improved principles which had been derived from recent experience. It was urgent to open the offensive as soon as possible, so as to anticipate the flood of American troops, and this limited time-span ruled out dealing the first blow in the damp fields of Flanders, which became accessible only in April. However there was reasonably suitable terrain on the line on either side of Saint-Quentin, even if it was obstructed somewhat by the craters of the old Somme battlefield. If all went well, the push on the chosen sector would split the British from the French, and enable the Germans to beat their enemies in detail by a planned sequence of blows which would incline them to peace. One disadvantage was that the attack on this southerly axis would hasten and facilitate the intervention of the French forces. In the event the Germans were able to deceive the enemy as to both the time and the place of the offensive, thanks to the skilful way the troops destined for the attack were held back from the front line, and the care with which they were concealed. About fifty of the German divisions were fully geared up for mobile warfare, though the shortage of equipment and horses meant that something less ambitious had to be demanded of the remaining divisions.[7]

On 21 March the first wave of thirty-seven divisions stormed forward on both banks of the Somme under cover of a barrage from 6,000 guns. The first blow was followed up on 6 April by an assault south of the Oise, and three days later by an attack on Armentières which led to the capture of the greater part of the Ypres Salient and the dominating Mount Kemmel. The German offensive cost the British about 300,000 men; 65,000 prisoners and 769 guns fell into our hands, and the British were forced to destroy many more guns and a great quantity of equipment. It amounted to the greatest success achieved on the Western Front since trench warfare had begun. The British had only 140,000 troops left as replacements, and they were forced to call off an offensive which they had planned in Palestine, as well as drawing two divisions from there and two more from Italy, and lowering the age of the call-up.

For a time the Germans retained the initiative, but they were never able to achieve the intended breakthrough.[8] The advance of the German infantry divisions meanwhile slowed down as it crossed the shell holes of the old Somme battlefield, and the enemy were gradually able to parry the attack, principally by having recourse to the army-level motor transport groups to throw forces against them.

We cannot declare categorically that the Germans *would* have accomplished the breakthrough if they too had possessed mobile troops, but it is a question which we cannot ignore when we look back on this episode. In view of the appalling condition of the roads behind the German front at the time, and the considerable volume of transport which was needed to sustain the infantry divisions and the artillery, it is very likely that only armoured units with full cross-country mobility would have had any chance of success; the opportunity was magnificent – of that there can be no doubt, for the enemy were heavily depleted and in a state of considerable disarray.

The end of April brought the failure of a further bid to break through in the direction of Amiens, and General Ludendorff now decided to attack across the Chemin-des-Dames towards Paris. This time it was the French who were the targets, and the British were able to make good use of the breathing-space. Forty-one German divisions and 1,158 batteries attacked the French on a frontage of 55 kilometres, and pushing rapidly forward the Germans reached the Marne between Château-Thierry and Dormans during the period 27 May – 1 June. Fifty thousand men and sixty guns were captured. The next attack was delivered at Noyon with the intention of relieving the pressure on the right flank of the German Seventh Army. This was unable to begin before 9 June, however, and it failed in the face of strong resistance by the French. The flank of Seventh Army was still vulnerable at Soissons, just like the flank of the First Army at Reims. Some anxiety was also aroused by the salients which the Germans had already made by their attacks in Flanders and at Amiens, for these would make them vulnerable if they were pushed on to the defensive. The German attack on the Chemin-des-Dames was a brilliant achievement in its own right, but it came to a halt at the Marne bridges and the forest of Villers-Cotterêts –

significantly enough when the French brought up tanks and rushed up infantry on trucks. This episode also happened to be the first appearance of the light Renault tank.

Ludendorff launched yet another attack, lest the initiative be wrested from him before the Americans intervened and our other enemies had the opportunity to recover. The Seventh and First Armies and elements of Third Army were to attack on each side of Reims so as to secure the German salient on the Marne. This blow was to be accompanied by a new offensive in Flanders. Employing the established tactics, 47 divisions and more than 2,000 batteries were to cross the Marne and take Reims, thus consolidating the German gains against the French. However on this occasion the attack was betrayed and the Germans failed to gain surprise. The French evacuated the eastern sector of the area of attack, and established themselves in difficult country west of Reims with the help of reinforcements and tanks. On 17 July Ludendorff ordered the offensive to be terminated, and the transfer of forces to Flanders began. The process was never completed.

The German tanks had taken an active part in the great offensive, but one cannot decide battles with just forty-five tanks. These were organized in detachments of five tanks each. The best way to make use of these scanty resources would have been to concentrate them as a combined force on some point where we needed to gain a rapid decision, and where the ground was reasonably favourable for the movement of tanks. But this was too much for the high command to swallow, and, ignoring the lessons of Cambrai, the Germans employed the tanks in small units, and sometimes even as single machines which were tied down to the infantry. Individually the tanks often did extremely well, but they could exercise no influence on the course of events.

Here we should add that the accelerated artillery bombardment was a characteristic of German tactics until the end of the war. Our experiences in the spring of 1918 indicated that it still worked well on the offensive, but when the Germans were on the defensive it proved of little avail against enemies who chose to attack with novel weapons and techniques.

From the beginning of June there were disturbing signs that the Allies were indeed beginning to fight in a new way, but it was on a small scale and did not receive the attention it deserved.

As early as the end of May the balance of forces had changed to such an extent that on the sector between the Oise and the Marne we had just 9½ divisions, largely exhausted by having been on the attack for several days, which were pitted against 11½ French divisions, most of them fresh. On the 31st the German Seventh Army attacked in the direction of Crépy-en-Valois and La Ferté-Milon, but it ran into stiff resistance, and in some locations the fresh enemy forces were able to launch successful counter-attacks. Attacking in several waves, the new light Renault tanks surprised 9th Jäger Division between Missy and Chaudun, and in addition gained the right flank of 14th Reserve Division. The German artillery did not spot the tanks until too late, and for a time there was something of a crisis. The French attack was beaten off, but the offensive capacity of the two German divisions had been crippled. On this day the progress of Seventh Army as a whole was modest, for the French were able to bring up fresh forces by truck. The Germans had to counter by committing their reserves.

Farther south the German 28th Reserve Division was pushed forward with the task of attacking generally north-west in the direction of Chouy. The result was a fair degree of confusion.

On 1 June the division proceeded to cross the Savières stream and establish itself on the west bank, but on its left wing it was repulsed from the village of Trosnes whose defence included three tanks; the whole attack now ground to a halt. The division was already in a dangerous situation when on 2 June it was sent forward on a broad front to take Villers-Cotterêts, and before the morning was out it ran into a French counter-attack which was spearheaded by tanks. The French were beaten off, thanks to the vigilance of the artillery, but the German losses were severe. A new German division came up on the left, which afforded some relief, and the 28th was able to assign its forces to a proper divisional reserve.

On 3 June the French committed still more tanks in their counter-attacks, which augmented the German losses. At 0530 28th Division attacked the sector Corcy–Vouty–Faverolle with 111th Reserve Jäger Regiment on the right and 110th Reserve Jäger Regiment on the left. At first the Germans made good progress under cover of the morning mist which eliminated the danger of French flanking fire. The respite came to an end when the Germans encountered heavy defensive fire from machine-guns, artillery and fighter aircraft, and finally at 0630 a further five tanks burst out of a wood north of Vouty and assailed 111th Reserve Jäger Regiment, breaking through the first lines of 3rd Battalion and forcing a number of the troops to give way. Two of the tanks were immobilized by Minenwerfer, but they continued to fire, and the remaining three machines turned north and drove back 2nd Battalion. Corcy was lost again. The three tanks now came under attack by the riflemen of 2nd and 3rd Battalions of the 111th, 1st and 3rd Battalions, 109th Reserve Jäger Regiment, and 3rd Battalion, 150th Jäger Regiment. The five battalions together were ultimately able to put the tanks out of action and take the crews prisoner. But just think for a moment! Five tanks with crews amounting to ten men had been able to reduce an entire division to disorder. In those 2½ hours 111th Reserve Jäger Regiment had lost a total of nineteen officers and 514 men, of whom two officers and 178 men were missing. There was now no question of 28th Reserve Division being able to resume the offensive. The 2nd Guards Division likewise suffered heavily from the attentions of the tanks, and on the same day Augusta Regiment lost twelve officers and nearly 600 men. Again on 4 June French tanks were able to stop the Germans exploiting another push.

In these combats the French tanks seem to have been sent into action with strictly limited aims, namely to prevent the Germans from penetrating the forest of Villers-Cotterêts and to secure a start-line for the intended offensive. These objectives were achieved. How well did the German infantry fare against them? It had been 1 and three-quarter years since the first ever appearance of tanks, on 15 September 1916, and six months had elapsed since Cambrai. What had actually been done to help the German infantry? What had they managed to learn? What could they reasonably be expected to do, given the heavy inroads that months of being on the offensive had made on their effectiveness?

The French had held back their tanks until the beginning of June, even though there had been loud calls for them to be sent in every time the Germans had broken through. But the commanders of the French tank forces had steeled themselves against such demands even before the German offensive had begun, and they were determined not to repeat the mistake the British had made in September 1916. They were adamant that they would commit the tanks only *en masse*, and only when all the divisions of one of their assault armies had been properly equipped with them. Because of manufacturing difficulties this programme did not progress as rapidly as had been hoped and by 1 May 1918, in addition to the sixteen groups of Schneiders and six of Saint-Chamonds, the French had at their disposal 216 *Chars légers* of which only 60 were immediately available for service. This was not very much, but by now the manufacturers and the tank forces at least had the satisfaction of knowing that the same people, who earlier had put the greatest obstacles in the way of tank development, were the ones who were now yelling the loudest for tanks to be flung at the enemy.

As we have seen, at the junior level of command the French were perfectly willing to work on the lines laid down by the tank officers. There were, however, a few exceptions, and we shall note them here. On 5 April a force of six tanks went into action to support a limited attack in the area of Sauvillers-Mongival; only one of the tanks reached the objective and the assault failed. Again, just six tanks were supposed to co-operate with a company of infantry to attack the park of Grivesnes; the tanks were successful in the short term, but the infantry did not follow up and the French were unable to hold the park after it had been captured. On 8 April, however, twelve tanks supported a successful attack on two woods north-west of Moreuil and Morisel. On 28 May another twelve tanks enabled the Americans to capture Cantigny and no machines were lost. The action at Chaudun on 31 May did not turn out so well. The French were striving to hold up the German advance, and six platoons of *Chars légers* together with units of the Division Marocaine attacked in an easterly direction straight off the line of march, and without preliminary reconnaissance or co-ordination with the infantry. The push was made in the middle of the day, across open ground, with no artillery support, no protection provided by mist, no aerial support and without any attempt on the part of the French infantry to follow up. The tanks were repulsed, whereupon they fell back to make contact with the infantry and then resumed the assault. The process was repeated over and over again and always with the same result. The tanks conquered an area twelve kilometres wide by two kilometres deep, but it was all lost again because the infantry were unable to follow up. The troops were exhausted, and the German machine-gunners kept up a deadly flanking fire against the attack, which had been delivered on a narrow frontage.

Over the following days see-saw combats of this kind were repeated along the Savières stream, and the eastern edge of the forest of Villers-Cotterêts. Ultimately nine companies of tanks played a significant part in halting what the French saw as the most dangerous of the German thrusts, aiming at Paris.

The French employed considerably greater tank forces against the German Gneisenau Offensive which erupted on 9 June from the area of Noyon in the general direction of Compiègne. On 10 June the push reached the line Méry–Belloy–Saint-Maur, with the leading units penetrating to the Aronde. The French resolved on a counter-attack for the morning of 11 June with four fresh divisions and four detachments or 'groupements' – two of them of Schneiders and two of Saint-Chamonds. The French armour moved up in secret during the night, and at 1000 a total of 160 tanks launched a surprise attack from the start-line Courcelles-Epayelles–Méry–Wacquemoulin. They accomplished their mission, which was to throw the Germans back into the valley of the Matz, and in the process they destroyed a large number of machine-guns and inflicted heavy casualties among the German infantry. But the tank forces also suffered severely (46 dead and 300 wounded, and 70 machines) whenever the German artillery had good observation or could fire over open sights. The attack had arrived very late, in full daylight, and because of this the French infantry were immediately spotted immediately and the German artillery and machine-gun fire prevented them from keeping up with the tanks. The tanks had to stay up front for a considerable time after they had reached their objectives, and French commentators believe that the delay, together with the lack of infantry support, accounts for the heavy losses. The area gained on this occasion measured eight kilometres wide by up to three kilometres deep.

As the number of tank units increased in the course of 1918, so the French were able to create tank regimental and brigade headquarters. The regiments were composed of a varying number of units according to circumstances, and the brigades comprised three regiments each.

From the middle of June the character of the fighting between the Marne and the Aisne began to change; the French now aimed to win a good base for their forthcoming offensive. Individual tank platoons and companies made a successful contribution in the course of these actions, but the losses remained heavy, and confirmed the high command in their conviction that the only appropriate way to employ tanks was in attacks by large numbers at a time. Yet again, on 16 and 17 July, three tank detachments of 502nd Regiment were used in the old style – beating off German attacks on the Marne south of Jaulgonne and Dormans; they lost fifteen of their machines. But while the gaze of the Germans in the wide Marne salient was still

1. British Mark I Tank, 1916.

2. British Mark V Tank, 1918.

3. French Schneider tank with infantry.

4. A knocked-out Schneider tank within the German lines on the Aisne, spring 1917.

5. A French Renault F.T. light tank, 1917.

7. A British Medium A (Whippet), 1918.

6. A French St. Chamond, 1917.

8. A British Mark II, 1929.

9. A German A7V tank, 1918.

10. A German LKII tank, 1918.

11. The British Vickers Independent, 1925/6.

12. French Chars 3C, 1928.

13. The French Renault, NC2, 1932.

14. American Cavalry Combat Car T2, 1931-3.

15. British Light Tank Mark II, 1931/2.

16. A French Renault UE with trailer.

17. The British Carden-Loyd amphibious tank, 1931.

18. The Russian Christie fast tank, 1933.

directed towards the south and south-east, a mighty storm had been brewing between the Marne and the Aisne, and it now burst upon them at a particularly inconvenient time. The French high command entrusted two armies – Mangin's Tenth Army north of the Ourcq and Degoutte's Sixth south of the river – to carry out an attack without artillery preparation, but with the support of a large number of tank units, on the model of Cambrai.

It was vital for the success of the offensive to keep the preparations secret. As far as the tanks were concerned, the tank commanders with Tenth Army received the order to move up their forces at midnight on 14 July; the tanks that were unable to travel by road were unloaded from trains at Pierrefonds, Villers-Cotterêts and Morienval on the 16th and 17th; the assembly of Sixth Army's tanks had been completed on the 15th. On the night of 17/18 July the tanks progressed to the start-line during a violent storm which covered their noise. The tables on pp. (98-9) indicate how the tanks were distributed among the assaulting divisions.

While the French command concentrated 490 tanks for the main attack, the considerable number of 180 machines remained inactive on the subordinate fronts. Sixth and Tenth Armies were to attack simultaneously and by surprise, with the aim of eliminating the 'Château-Thierry Pocket', or at least making the nodal point of Soissons unusable by the Germans. At the same time as Sixth and Tenth Armies attacked from west to east, Fifth Army south of the Vesle was supposed to thrust in the opposite direction towards Arcis-le-Ponsart, although the necessary order could be given only after it had become clear that the German attack of 15 July had failed.

French Tenth Army was to attack at 0535 on 18 July behind a creeping barrage. The first phase line ran from Berzy-le-Sec by way of Chaudun to Vierzy. After the objective had been reached II Cavalry Corps would exploit the success, with its 4th Cavalry Division advancing from Taillefontaine (twelve kilometres behind the front line) by way of Chaudun and Hartennes to Fère-en-Tardenois, and its 6th Cavalry Division pushing from Vaumoise (eighteen kilometres behind the front) through Vertes-Feuilles, Vierzy and Saint-Rémy to Oulchy-le-Château. Second Cavalry Division was to follow the 4th as corps reserve. Fighter aircraft were put at the disposal of the offensive, and six infantry battalions, together with engineers, were held ready on trucks at Mortefontaine and Villers-Cotterêts. Sixth Army was to set out at the same time as Tenth.

At 0535 the French artillery opened a brief but intense barrage and the tanks and infantry stormed forward. The approach was screened by a light mist, and the Germans were taken by surprise. As early as 0830 Tenth Army alone had effected penetrations of more than three kilometres on a frontage of twelve, and by noon it had driven six kilometres into the German defences along the decisive axis. In the afternoon the French confined themselves to strictly limited gains, and not until evening did the arrival of fresh tanks give the attack a new impetus, which carried the French two kilometres beyond Vierzy. Tenth Army had made penetrations averaging five to six kilometres along a frontage of fifteen kilometres, with local gains of nine kilometres; the advance of Sixth Army farther south reached a depth of about five kilometres.

The offensive hit ten divisions of German Ninth and Seventh Armies, with seven divisions ready in support. Their divisional frontages measured 4½ to five kilometres, as against the 2-kilometre frontages of the French assault divisions. The German troops were in a thoroughly bad way; their heavy losses in the earlier offensives had not been made up, there were few strong positions available, and their supplies were inadequate. Altogether the combat effectiveness and endurance of the troops were not what they had been, and when the French surprise attack arrived most of the German infantry were wiped out in their positions and the artillery was lost.

And yet the attack was virtually over by 0840. How was it possible for the offensive to run out of steam and the French gains to prove so insignificant? How do we account for an episode like the one on the German right in Army Group Staabs' sector, where 241st Jäger Division, after its southern wing had been annihilated by tanks, was able to withdraw half its complement from its hitherto intact front line, and withdraw virtually unmolested by way of the Aisne valley towards Soissons? Why were there complete intermissions in the French artillery fire from noon onwards? How

Sixth Army

Corps	1st Line Divisions	2nd Line Divisions	Reser.	Available		
				Guns	Tanks	Aircraft
II.	33rd	-	-	144 Field	-	40
	Half 4th U.S.	-	-	108 Heavy	-	-
	2nd	-	-	-	45	-
		63. (Army res.)	-	-	30	-
	47th	-	-	-	45 + 12	-
VII.	Half 4th U.S.	-	-	36 Field	-	30
	164th	-	-	84 Heavy	15	-
I U.S.	167th.	-	-	84 Field	-	30
	26th U.S.	-	-	84 Heavy	-	462 (Army)
	7	1	-	588	147	562

Elsewhere were deployed :

Ninth Army .... 90 Chars légers
I Cavalry Corps with Fifth Army .... 45 Chars légers
and elsewhere probably a further .... 45 Chars légers

Approx. total 180 Chars légers

was the wreckage of 11th Bavarian Jäger Division able to occupy and hold the ridge west of Vauxbuin in the face of the victorious enemy? For a time the division in question had only two battalions left! In the afternoon reinforcements brought the complement up to seven, and in the evening to as many as nine! In front the Germans could identify French forces regrouping for a new attack, as well as artillery changing position, and tanks and even cavalry. And yet the Germans were given a night in which to put their units in order and prepare an obstinate defence.

It was the same story with the neighbouring Army Group Watter to the south. When the French artillery opened up the German supporting troops were put on alert, and the German artillery laid down a barrage. Only the two right-hand divisions of the army group were struck by the main thrust of the French attack. On this sector the French seized Missy as early as

Tenth Army

Corps	1st Line Divisions	2nd Line Divisions	Reser.	Available		
				Guns	Tanks	Air-craft
I.	162nd 11th 153rd	- 72nd	- - -	228 Field 188 Heavy -	- - 27 Schneiders	40
II.	1st U.S. Moroccan 2nd U.S.	69th 58th (Army res.)	- - -	276 Field 172 Heavy (incl 69th & 58th div. arty.)	60 St.Chamonds 48 Schneiders 48 Schneiders	50
III.	38th 48th	19th 1st (Army res.)	- -	216 Field 112 Heavy (incl 19th & 1st div. arty.)	30 St.Chamonds -	50
XI.	128th 41st	5th	- -	114 Field 128 Heavy	- -	40
II. Cavalry			2nd Cav. Div. 4th Cav. Div. 6th Cav. Div. 6 inf. bns. on trucks		1st, 2nd, 3rd Renault bns. 130 Chars légers	301 (Army)
Total	10	6	-	1545	343	481

0820. The artillery of 42nd Jäger Division offered what resistance it could against the French tanks, which could hardly be seen in the high corn, but it had to give way, and by 0830 the Germans had lost all the guns they had deployed to the west of the Chaudun–Missy position. Nevertheless it was here that the Germans were able to establish an organized defence. In the sector of 14th Reserve Division (with which three regiments of 46th Reserve Division were also fighting) the Germans were surprised to see that the French attack had ignored the deeply cut valley of the Savières, with its wooded banks; the French were content to keep the gorge under a suppressive fire while they directed the main thrust of their attack along the heights to the north and south. The reason was that the Savières ravine offered little opportunity for tanks to deploy or operate effectively, and the French instead intended to capture it by a double turning movement.

When the Germans made their appreciation of the terrain they had failed to take into account the nature of the French offensive tactics, an oversight which probably enabled the French to achieve a still greater degree of surprise, and facilitated their breakthrough at Vauxcastille. Thus 159th Jäger Division held out gallantly in the Savières valley until it was wiped out by the double envelopment. When the French broke through at 0600 only one officer, four NCOs and six men escaped from 53rd Reserve Jäger Regiment which had adjoined to the left. The parent 14th Division lost its artillery. It took the last German reserves, including even some companies of Landsturm, to occupy Vierzy at 0730.

Fourteenth Division's left-hand neighbour was 115th Jäger Division, which was able to beat off the French attack, except for one minor incursion. The reason? The enemy had no tanks. However the division was being embraced by a double envelopment, and it had to be pulled out in the evening.

By 0800 the German high command had a reasonably comprehensive view of the critical situation of the front, and ordered its forces to occupy and hold a position running from Chaudun through Vierzy to Mauloy. An additional regiment of infantry was assigned to every division for this purpose, though no artillery was available. It is instructive to investigate just how effective these measures turned out to be, especially since the enemy tank attack hit the Army Group Watter with particular force and speed:

(a) The 109th Grenadier Regiment was put at the disposal of 42nd Jäger Division, but its two battalions arrived too late, for at 0930 the enemy had already taken the Chaudun position by a massive tank attack; nevertheless the tanks were prevented from advancing any further by the regiment's assigned artillery, namely 2nd Battery, 14th Field Regiment.

(b) Fourteenth Reserve Division was assigned 40th Fusilier Regiment, which set out at 0845 from Visigneux for Léchelle and braved intense artillery fire to reach the ridge south-west of Chaudun just before the French did; here it was able to hold out with the help of its accompanying battery of artillery (3rd, 14th Field Artillery Regiment) and two anti-tank platoons. The enemy attacks fell away from 1330, and the Germans were able to put themselves in order and restore a measure of cohesion among their units. Only one battery remained out of those which had been present at the beginning of the action, but the lull in the fighting enabled the Germans to increase the number to five, accompanied by supporting batteries of 40th Jäger Regiment and 16th Reserve Jäger Regiment.

(c) Second Reserve Jäger Regiment had been placed under 155th Jäger Division. It had already been positioned immediately behind the front line in the Mauloy wood as corps reserve. As early as 0730 two of its battalions were thrown into the battle, and the third was made ready to intervene as well. It is notable that this division was in place at the right time, and behind the only division that had not been hit by the tanks. The only artillery lost was a single sub-group.

Army Group Watter still had a corps reserve available in the shape of an infantry regiment at Villemontoire and another at Tigny. At 1400 group headquarters ordered the main supply train and all the spare vehicles to be transferred to the north bank of the Aisne, a move which was accomplished completely undisturbed. The 42nd Jäger Regiment was able to beat off isolated enemy pushes in the afternoon and evening, and the German artillery knocked out a number of tanks in the process. In contrast an attack on an altogether larger scale was aimed at 14th Reserve Division at 2030, and the intervention of fresh tank forces by way of Vierzy brought about a French success in the way already mentioned. How was it that the attack of the French tank reserves came on the scene so late? After all, it was only fourteen kilometres from their assembly area between Puiseux and Fleury.

Army Group Winkler was in action to the left of the Watter group. The defensive tactics were the same in both cases, and on this sector too the enemy avoided difficult features of the ground that would have held up the attack, in this case the Buisson de Cresnes, where 40th Jäger Division (the northernmost division of the group) was spared the attention of the tanks, and was able to hold out for a reasonably long time. The French had an initial success against 10th Bavarian Jäger Division, after their tanks went into action on 0930, and they were able to penetrate about 3½ kilometres in the direction of Neuilly-Saint-Front. Then, however, they were checked by the initiative of a number of junior commanders. The German success is all the more remarkable since the main French blow, complete with 132 tanks, was aimed at this division. How did this come about?

The answer lies in the way the tanks were incorporated with the infantry. Tenth Bavarian Jäger Division was attacked by two lines of French divisions – 2nd and 47th Divisions in the first line, and 63rd in the second. It seems that the French had not intended that 63rd Division come into action on the first day, and yet it was assigned thirty tanks, which were consequently not available for the attack on 18 July. Of the remaining 102 tanks, 45 were placed under 2nd Division, and 57 under 47th; the divisions in turn subdivided the tanks among their waves of infantry. The first shock was therefore delivered with only part of the total tank forces. There followed a pause in the attack, while the French were shifting their artillery, which permitted the Germans to restore their units to order. The French did not put in a fresh attack until 1745, and it failed. Whether through lack of perception, or lack of boldness, the French had failed to use their armour to maintain the impetus at the critical moment when their guns were changing position.

The German formation next in line was 78th Reserve Division. It did not come under such direct pressure from the tanks as the others, but its northern flank became increasingly vulnerable and it too had to pull back, losing some of its batteries in the process.

The Winkler Group received a significant reinforcement in the shape of 51st Reserve Division, which had been ordered to march north-west from the area of Beuvardes as early as 0720. Not long afterwards, at 1100, the first elements reached Armentières-sur-Ourcq, south-west of Oulchy-le-Château, eleven kilometres from the current front line. Just as had happened farther north, the defenders had succeeded in moving up strong reserves in a matter of hours, which shows just how little time is available for an attacking force to accomplish a genuine breakthrough, even after it has brought off a complete surprise. And that was back in 1918, when the Germans had to bring up most of their reserves on foot! On 18 July the only German formation to be moved by trucks was 10th Jäger Division, which was transported from Beuvardes to Nampteuil-sous-Muret, Muret-et-Crouttes and Droizy, and went into action on the same evening. In the present period of motorized and airmobile reserves we must attach still greater importance to maintaining the speed of the attack.

It remains only to mention the Corps Group Schoeler, which was deployed between Saint-Gengoulph and the Marne at Château-Thierry. It was struck by the offensive only on its far right wing, where it lost Courchamps.

By the evening of 18 July the French had made a successful break-in along a sector of forty kilometres, the entire frontage of their attack. A considerable number of German divisions were on the verge of disorder and others had been badly mauled.

Why was it that the break-in did not develop into an actual break-through? The explanation is of great relevance for the future employment, and consequently the organization of armoured forces. Among other things our analysis must address the following issues:

(a) how the forces, and especially the tanks, were committed to the attack;
(b) the relevant tank tactics;
(c) the composition and employment of the reserves.

We shall examine these points one by one.

(a) The French high command adjudged correctly that the German forces in the Marne salient, though numerically large, were badly placed, not least on account of the vulnerability of the main lines of communication, which ran through Soissons. The main blow was therefore going to be struck from the west from the forest of Villers-Cotterêts, and the task was entrusted to

Tenth Army. The high command further decided to depart from their earlier practice, and mount a surprise attack on the model of Cambrai, employing tanks *en masse*. Having determined on a thrust from Villers-Cotterêts, the French should then have concentrated all their available offensive capacity, by which we mean primarily their tanks and aircraft, on the axis in question. Moreover Fifth, Ninth and Sixth Armies should have done without tanks, so as to bolster the armour of Tenth Army. Lack of room should not have been an issue – the French did not have such an enormous number of tanks that Tenth Army could not have found space for them, especially if the French had concentrated only as many of the conventional weapons as were needed for an effective attack in the circumstances of positional warfare. Again, the terrain to the north and south of the Ourcq presented the same degree of difficulty, and there was no particular reason to divide the tanks among two armies. It is worth examining how the tanks could have been employed, if all the units had been committed with Tenth Army, in other words north of the Ourcq. Here we have to pay special attention to the deeply cut ravines which run south from the Aisne, namely those of Pernant, Saconin-et-Breuil and the valley of the Crise with its branches. The greater part of the tanks would have had to be directed south of the first two gorges, in the general direction of Grand-Rozoy and Hartennes.

(b) From their earlier tank actions the French concluded that only the closest co-operation with the infantry would turn an armoured attack to real advantage, promoting the conduct of the battle as a whole. Tanks were accordingly assigned to each line of infantry, and only the three newest and fastest tank detachments were held aside as army reserve. Likewise to the French way of thinking it was possible to launch a successful attack without a preparatory bombardment, but not without continuing artillery support, which could be provided only by the creeping barrage. When the wall of fire reached its maximum range there was only one thing to do – for the batteries to go forward. But moving great masses of artillery was a matter of several hours, especially in the era of horse-drawn artillery, and while this was going on the assault forces – the infantry and tanks – had to stop and wait just when they had the bit between their teeth. This was generally in open ground where men and machines were exposed to observation and an increasingly lethal fire from the defenders. So it was that the attackers gained surprise, only to throw it away. There are also signs that the creeping barrage was applied in a rigid way, since it did not touch a number of readily identifiable centres of resistance and strongpoints which were sited in difficult terrain. These could not be tackled by the tanks, and the result was that the attacking infantry became hung up on the flanking fire. Experiences of this kind will be repeated as long as the tank attack is coupled with unprotected infantry and horse-drawn artillery.

The practice of subordinating tanks to each line of infantry also frustrated any attempt to exploit the success of the first wave of armour in a speedy and energetic way. The senior tank commanders were banished from the scene of action, and degraded as 'advisers' to the higher

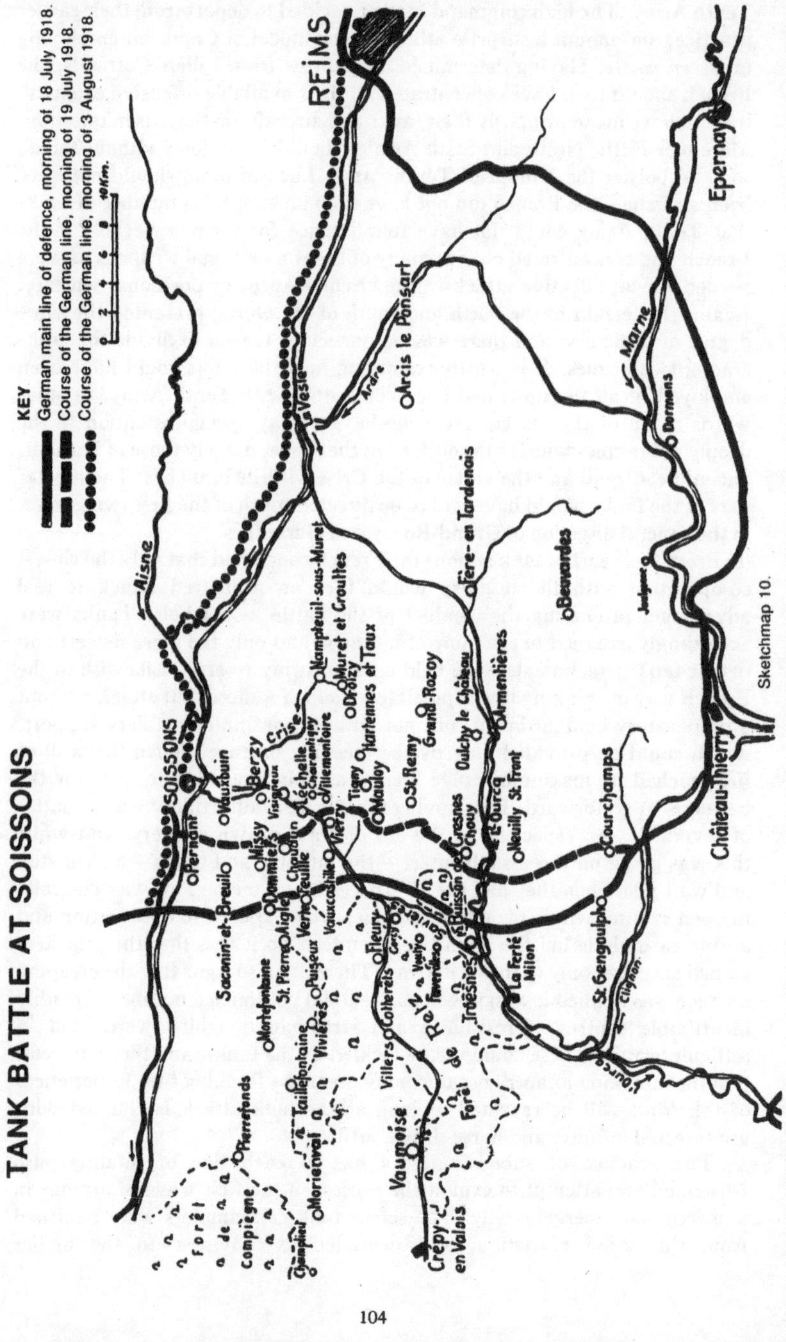

TANK BATTLE AT SOISSONS

KEY
German main line of defence, morning of 18 July 1918.
Course of the German line, morning of 19 July 1918.
Course of the German line, morning of 3 August 1918.

Sketchmap 10.

headquarters, where they had a thankless task. During the battle they made themselves unpopular because they seemed to disturb the smooth flow of operational thought with their tactical demands and technical objections. Afterwards they asked themselves where they had gone wrong, and set about the dismal business of patching up the remnant of their once proud squadrons.

(c) The reserves of Tenth Army most directly concerned with the events of 18 July consisted of:

Four infantry divisions, of which two were behind XX Corps and two behind XXX Corps;

Three cavalry divisions, of which two (at Taillefontaine) were behind XX Corps, and two (at Vaumoise) behind XXX Corps;

Three battalions on trucks at Mortefontaine behind XXX Corps;

Three battalions on trucks at Vivières behind XXX Corps;

Three tank detachments between Fleury and Puiseux behind XXX Corps.

By the standards of the time the French command believed it had made excellent provision for mobile and fast-moving reserves. It was right. All the same, the business of organizing and deploying the reserves seems to have occasioned a certain amount of friction, as we shall now see.

As early as 0815 the army commander ordered the Cavalry Corps to move up its various divisions. The formations duly set off, but they made slow progress along the roads, which were encumbered with other forces, and the result was to jam all the routes completely. At 1500 4th Cavalry Division reached Dommiers and Saint-Pierre-Aigle, and 6th Cavalry Division arrived west of the farm of Vertes Feuilles. Only now did the truck-borne infantry leave their assembly areas at Mortefontaine and Vivières, just seven or eight kilometres behind the old French front line. It soon became evident that there could be no question of making any further progress on horseback, and the French had to be content with sending a few detachments of riflemen towards Vierzy and the area to the south, to be inserted among the infantry who were already in action there. Nothing more was heard of the truck-borne infantry or of 2nd Cavalry Division; it seems that they were simply blocking the roads.

We return to the three separate tank detachments, which consisted of the newest and fastest machines, as already mentioned. Just after 1000 they received the order to send two of the detachments in the wake of XX Corps, and the third after XXX Corps, though with the restriction that they were to go into action only after the first line could make no further headway. At 2000 the 1st Detachment supported the action of 2nd American Division, advancing from Vauxcastille against the line Hartennes–Taux, and they were able to push the infantry along with them for three or four kilometres. Otherwise only a single company of the 2nd Detachment got into the fighting, at Léchelle; the results are unknown. The rest of the tanks failed to get into action at all.

It would have been quite possible to combine the three detachments of *Chars légers* attached to 501st Regiment, as a single formation under their

TANK BATTLE AT SOISSONS, 18 JULY 1918

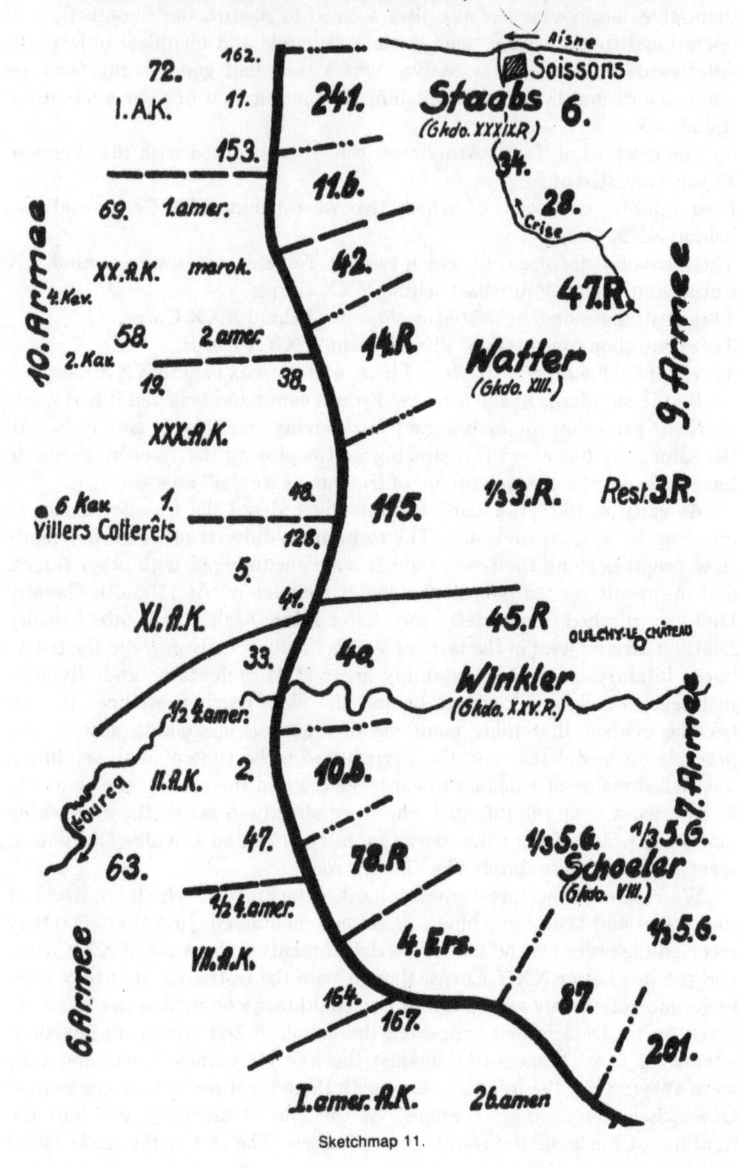

Sketchmap 11.

regimental commander, and push them forward in a simultaneous attack along the axis where the advance was making the best progress, in other words in the direction of Hartennes. The necessary orders should have been given with the least delay, and the regimental commanders and low-ranking officers should have been subject to as few restrictions as possible, so as to encourage displays of initiative. As things went, even the most dashing leadership would have been paralysed by the way the attack was tied to the progress of the creeping barrage and the rate at which the batteries could change their positions.

Thus the tank reserves should have been held immediately to the rear of the foremost lines, with the truck-borne infantry and engineers immediately behind the tank reserves and ideally under their command. The mobile infantry and engineers in their turn ought to have been pushed as far forward behind the tanks as the enemy fire permitted. They would probably have done this briskly enough, since they were in a physically fresh condition, unlike the infantry in the first line, who were exhausted by the fighting during the morning.

The intervention of horsed cavalry should have been considered only after the breakthrough had succeeded, when open country had been reached, and where there were no barbed wire entanglements, no trenches and above all no machine-guns to interfere with the speed and cross-country mobility of the mounted arm.

Three divisions were moved up from the infantry reserves on 18 July, and their place was taken by a single front-line division and the Cavalry Corps. Orders were given to bring up seven fresh divisions – mostly on trucks – for the 19th. Altogether on the first day of the offensive the French had taken 12,000 prisoners and 250 guns.

The events of the 18th are worth reviewing in perspective, since it was the first time that the French had combined the mass use of tanks with surprise.

The Germans had suffered a reverse of major proportions, and they at once got down to investigating the reasons in detail. They agreed on one thing – the French had achieved a total surprise, catching off guard not only the German troops but also to a considerable extent the high command. After no French attack had materialized south of Soissons on 15 July, the Germans were probably inclined to believe that their own push on Epernay had tied the French down. This was a miscalculation. But this miscalculation would have been attended by no further effects – and indeed the French would probably have not attempted a surprise attack at all – if the French had not had at their disposal the potent new weapon of the tank. Soissons was the first battle in which tanks were employed in anything like sufficient numbers, and with operational and not merely local objectives in mind. If 400 tanks had gone into action at Cambrai in 1917, now a full 500 were committed by the French at Soissons. The frontage was admittedly narrower than at Cambrai, and the blow was consequently no heavier; however the tank is the embodiment of the power of the offensive, and employed as it was on both occasions with the element of surprise, it took a heavy toll of the defenders in terms of blood and morale.

Returning to an earlier issue, it is depressing for us to have to relate that eight months after Cambrai the German infantry and artillery still had no anti-tank weapons; perhaps more depressing still is the failure to evolve any appropriate tactics to deal with this new threat. For some time yet the Germans continued to pay a heavy price in blood before they identified the root cause of their defeats, and recognized the need for an effective defence against this new weapon which had appeared in the armoury of the offensive. Unfortunately the Germans came around to the idea too late for it to have any influence during the war, and they forgot it again while they were labouring under the restrictions of the Versailles *Diktat*. At Soissons on 18 July the infantry never came to terms with the armoured attack at all, and it was only in the afternoon that the artillery began to hit back at the tanks from new and more intelligently chosen sites. This episode (and it was not going to be the last) should have shattered the dream world of the people who had dismissed the surprise attack by tanks as a 'one-off' weapon.

There were those who maintained that the German defence had failed because the front-line strength of the infantry was low and the troops were psychologically exhausted. Considering the way the enemy attacked, however, stronger combatant forces would probably not have failed to avert the catastrophe, but would have been likely to increase the losses. The men were undeniably in poor physical shape because of influenza and inadequate rations, but our brief narrative has revealed episodes of heroic endurance and lively initiative which give the lie to any allegations of flagging morale. Our admiration will only be increased by a detailed study of the combat record of individual divisions and regiments.

Careful analysis of the fighting on 18 July reveals three fundamentally important reasons for the German failure:
(a) successful surprise on the part of the French;
(b) the striking force of the French tanks, which made that surprise a reality;
(c) the fact that the German artillery, but especially the German infantry, had no effective weapons or tactics with which to combat tanks.

Over the following days the German defeat assumed still larger dimensions, extending beyond the tactical level and well into the operational sphere. The reason was that the French drive on Soissons posed a dire threat to the communications of the German armies in the Marne salient, compelling the German command to evacuate the recent gains on the south bank of the Marne, and withdraw the front line to behind the Vesle. Ten German divisions were disbanded, as a result of the losses in dead, wounded and prisoners, and the high command had to abandon the intended 'Hagen Offensive' in Flanders. The Germans went on to the defensive along the entire Western Front and the initiative passed to the enemy.[9]

From the perspective of the French, we have to ask why the attackers did not make a breakthrough on the first day of the offensive and cut off the Marne salient there and then. We have already noted that the French could have tied in the available tank units more closely on the decisive sector of the offensive, namely on the frontage of Tenth Army – a measure which would have given that army much greater striking power on the vital axis of the plateau south of Soissons. But something more was required to accelerate the tempo of the attack as a whole, and enable the French to exploit their surprise more effectively. The trouble was that the pace of the armoured attack was dictated by the progress of the other arms – the infantry, as they moved slowly forward and came under the fire of hidden machine-guns, and the French artillery, with its systematic creeping barrage and its changes of position – which took hours with its horse-drawn transport. Meanwhile the tanks had to wait within range of steadily increasing fire from the defenders. As long as this state of affairs continued the French could bring off break-ins, but not proper breakthroughs, and the defenders were always able to consolidate a new line. This meant that the French had to mount an entirely new attack – something which could not be improvised overnight – and the element of surprise vanished completely. As things were managed in 1918, there could be no prospect of armoured support either, for the tank forces were exhausted by the first day of action. Another consequence of incorporating the tanks so completely with the infantry formations was that only a proportion of the machines ever got into combat; in the case of Tenth Army we are talking about 223 tanks of the 343 available; 120 were tied down with the rearmost infantry lines and reserves, and were completely ineffective. The French undoubtedly made tactical gains in the fighting on 18 July, but they were far from exploiting to the full the potential of the new weapon, which derived from its speed, armoured protection and firepower.

The remaining French pushes represented nothing fundamentally new. Despite the loss of 102 tanks Tenth Army still had 241 machines available on 19 July, but only 105 were in action. Thirty-two tanks were engaged on 20 July, 100 on 21 July, and 82 on 23 July. In the period between 18 and 20 July the Tenth Army lost altogether 248 tanks, at least 112 of them to artillery fire. 'Irrespective of range the artillery piece proved to be the main enemy of the tank. Success in combat hung essentially on protection against the enemy guns.' (Dutil, *Les Chars d'assaut*, Paris 1919.)

Let us now take stock of some developments in August 1918. At the beginning of that month the French armnoured units consisted of ten battalions of *Chars légers* and eight groupements of medium tanks (Schneiders and Saint-Chamonds).

The Germans, after they had beaten off the French attempt at breakthrough, withdrew their front line sector by sector behind the Vesle. They arrived there on 2 August, having suffered heavily in some courageous rearguard actions.

The German high command hoped that the hostile alliance had exhausted its offensive capacity, and that the severely depleted German divisions would be granted a little respite in the immediate future. It is not clear from the historical record whether the Germans drew the appropriate lessons from the Battle of Soissons, and followed their useful habit of communicating them without delay to the other sectors of the front. At all events there is no evidence of a fundamental change in tactics, least of all in the artillery. Just as before, defensive fire was categorized as 'long-range barrage', 'short-range barrage', and 'annihilation fire' on identified or suspected concentrations. It should have been demonstrated at Cambrai and only very recently at Soissons that defensive fire of this kind was totally useless against surprise attack with tanks. These two actions, and the successful defensive battle on the Aisne, showed that the real tank-killers were direct aimed fire by batteries, the direct, individually aimed round, and on occasion well-observed fire from heavy batteries. And yet at the beginning of August there was still no fundamental change in the siting and modes of fire of the artillery.

It was the same story with Second Army, which held the westernmost projection of the German front at Amiens. In August 1918 all the front-line divisions were compressed into a single deep deployment, despite their weak combat strength. There was an almost complete lack of strongpoints, and the artillery was sited where it was useless for anti-tank defence. The result was that the enemy tanks were able to attack without being bothered in any significant way by defensive positions, artificial or natural obstacles or artillery fire.

We should in no way blame the German infantry for the failure to strengthen their positions after the great spring offensive terminated on 24 April. It was partly a question of the Germans clinging to the hope of resuming the offensive some day or another, and partly the result of the exhaustion and numerical weakness of most of Second Army's front-line troops. Dominating everything else, however, was the ceaseless rain of shells which made labour on the trenches so extraordinarily difficult and costly, and destroyed much of the work as soon as it had been finished. In addition long sectors of the front were lost in the endless fighting, so that the enemy actually reaped the benefit of all the effort the Germans had put into building strongpoints. Villers-Bretonneux and Hamel are cases in point.

On 24 April 1918 Villers-Bretonneux was the scene of the first action of tank against tank, and we shall return to it later in that context. We will just note here that the appearance of German tanks in the field had the immediate effect of hastening the dispatch of further British tanks to France. The British acted on the principle that the finest tanks in the world were unable to withstand other tanks, and that the only way to meet an armoured attack was to have a greater number of tanks of one's own.[10]

Sixty new tanks were being delivered to the British every week, and the British attack on Hamel on 2 July should have given the Germans an opportunity to evaluate the performance of the enemy machines by this stage in the war. The tanks in question were the Mark Vs, and at the express wish of General Elles they received their baptism of fire in an attack with a limited objective, namely the capture of Hamel by the Australians.

The assaulting infantry and the crews of the new tanks came to know and trust one another in the course of joint training, and Colonel Fuller worked out tactical details with some care. There was no preparatory bombardment, and at 0410 three Australian brigades attacked with the support of sixty tanks and under cover of a creeping barrage of smoke and high-explosives. The tanks started out 1,000 metres behind the leading line of infantry, but they rapidly overhauled the foot soldiers and hastened to their objectives. With the advantage of surprise, the assault crashed through the German lines on the entire 4-kilometre frontage of the attack, wiping out most of the defenders, destroying 200 machine-guns and taking 1,500 prisoners. The Australians had 672 casualties and sixteen of the tank personnel were wounded; six tanks were slightly damaged. Just half an hour after the objective had been reached, four supply tanks laden with twenty-five tons of engineer equipment arrived immediately behind the new front line. The action at Hamel was possibly of minor importance in its own right, but it encouraged the British command to devise a new tank battle on the grand scale.[11] Had the German leaders learned any corresponding lessons for the defence? Evidently not.

On 23 July three French divisions attacked the bridgehead west of Moreuil with the support of a British tank battalion. Losses on this occasion were heavy, since, contrary to plan, the attack opened only some time after first light; fifteen out of thirty-five tanks were damaged, and fifty-four officers and men of the battalion were killed or wounded. Nevertheless the objectives were taken, together with 1,800 men, 275 machine-guns and a number of artillery pieces.

These successes confirmed the confidence of the British high command in the striking-power of their armoured forces, and it went ahead with preparing its mighty blow. For weeks now the British had posssessed complete mastery of the air, which gave them a detailed knowledge of the German positions, and further information from prisoners and other sources left them in no doubt as to the state and deployment of the enemy forces. So it was that nine German divisions had to face eight divisions of British and five of French. Three British and two French infantry divisions and one British cavalry division came up behind as reserves; the German reserve divisions numbered five. All the Allied formations were rested and fully up to strength, which was the case with only two of the German divisions.

And yet the considerable numerical superiority of the British, Australian, Canadian and French infantry, and of the enemy guns and ammunition would not have been enough in themselves to guarantee a breakthrough of the German front, if the Allies had been forced to depend on artillery fire and infantry assault alone. German infantry and machine-gunners had already beaten off other attacks. Nor can we ascribe our misfortune to the fog which shrouded the battlefield on the morning of 8 August – after all it had been foggy on the Somme and at Ypres, without the enemy turning it to tactical advantage! No, none of this explains how we were overtaken by our sudden, appalling Black Day. Our army was battle-tested, if no longer fully battleworthy. Our infantry were as determined as ever to stick it out; and accounts of the time speak of hardships, but also of a spirit of soldierly defiance. Posterity will do an injustice to the self-sacrifice and courage of many thousands of our soldiers if it alleges any collapse of nerve, panic at the sight of tanks, or dereliction of duty in the presence of the enemy. If a few soldiers failed, they cannot derogate from the heroic – and for that very reason – tragic, endurance of the overwhelming majority of the combatants. It is in this perspective that we will investigate the events of 8 August 1918.

This was the third time that the Germans had been confronted with a battle on the Cambrai model, and for the third time they allowed themselves to be taken by surprise. The enemy deployed on the nights immediately before the storm, moving up in strict silence and observing the most elaborate precautions. Diversionary convoys, traffic and activity all served to conceal the assembly of the designated forces, namely the Canadian Corps (which had a reputation for offensive-mindedness) and the Tank Corps.

The distribution of the tanks is given in our table, which presents the deployment of trops from north to south.

On the night of 6/7 August the Tank Corps assembled three or four kilometres behind the front line, and on the night of 7/8th moved forward to its start-line about one kilometre from the front. From the way the tanks

British Army

Corps	Divisions		Res.	Tank form'ns and units		Number of Tanks	
	1st Line	2nd Line		Bdg.	Btn.	1st Line	2nd Line
III.	12th	-	-	-	-	-	-
	18th	-	-		10th	24	-
	58th	-	-			12	-
Australian	3rd		-	5th		24	-
		4th	-		2nd 8th	-	54
	2nd		-		13th 15th	24	-
	-	5th	-			-	42
Canadian	2nd		-	4th	14th	36	-
	1st		-		4th	36	-
		4th	-		1st	-	36
	3rd		-		5th	36	-
Cavalry	-	-	1st	3	3	48	3rd Line behind the Canadian Corps
	-	-	2nd		6	48	
	-	-	3rd		-	-7	

were assigned to the assaulting divisions we can identify the way a number of axes of effort were formed among the Australian and Canadian forces. Within the divisions, however, the allocation of the tanks indicates that, just as had happened at Cambrai and Soissons, the tanks were tied in closely with the lines of infantry; the two most modern and fastest-moving battalions, namely the 3rd and 6th, which were equipped with Whippets, were placed under the Cavalry Corps, which was three divisions strong, and deployed between Cachy and Amiens to exploit the success and complete the breakthrough. The attack was timed to begin at 0500, and the artillery was to be employed partly to put down a creeping barrage of smoke and high-explosives in front of the assaulting infantry and tanks, and partly to suppress the German batteries and other long-range targets. The 500 aircraft were likewise incorporated in the plan of attack, whether directing fire and carrying out combat reconnaissance, or attacking targets in depth.

The first targets lay 1½ to three kilometres deep in the German defences and they were to be reached by 0720, but the German batteries facing British III Corps remained completely outside these initial objectives; the Australian attack was to reach only the most advanced German batteries. The Canadian attack was to penetrate considerably deeper, embracing most of the German gun positions on their sector, but on the French front again only a few batteries came under attack. While most of the German batteries remained untouched, British III Corps push north of the Somme was to be delayed for an hour, and its blow south of the river for two hours, so as to permit the rearward waves to come up and continue the attack, and for the artillery to finish changing positions. After this pause – under the muzzles of the German guns, as already indicated! – the creeping barrage ceased and the attack was to be supported by artillery acting in accordance with the procedures for mobile warfare.

The second objective took in the German batteries on the whole 30-kilometre frontage of the attack, while the third phase line fell just short of the quarters of the German reserve divisions, even though their location must surely have been known to the enemy. When the attack resumed at 0920 it was to continue without a pause. This was when the Cavalry Corps was supposed to advance with one division to the north of the Luce and another to the south, then overtake the infantry, continue to the third objective, hold it until the infantry came up, and finally push to the final objective, the railway between Chaulnes and Roye.

The French opened fire simultaneously with the British at 0520, but waited for three-quarters of an hour for the bombardment to take effect before the attack of the first line, which consisted of three divisions of infantry without tanks. Only after the French had taken the dominating heights west of the Avre would 153rd Division with the two battalions of *Chars légers* proceed through the first line of infantry and drive in the direction of Hangest-en-Santerre. For a considerable time, therefore, the French were in danger of making slower progress than their Canadian neighbours, which left the latter vulnerable to flanking fire. The German artillery made the best of this opportunity, especially to shoot up the tanks on the right wing of the Canadians.

Once again the enemy had made the mistake of tying down the tanks to the infantry and artillery, and on this occasion they extended it to their most promising weapons, the two most mobile battalions with their Whippet tanks, which were linked with cavalry which had no place on the modern battlefield. Confined within this rigid framework, did there remain any chance of the offensive achieving a triumphal breakthrough? Hardly. Nevertheless the attack put the Germans through some alarmingly dangerous experiences, as we shall now outline.

The Allied armies went into battle confident of victory, while the Germans awaited their fate anxiously from day to day. On 6 August a German aircraft reported one hundred tanks on the move from Ailly-sur-Noye to Morisel. The Germans took no particular fright. On 7 August a chance hit blew up twenty-four supply tanks laden with ammunition and fuel in an orchard at Villers-Bretonneux. Again it aroused no suspicion. At 0520 on 8 August the enemy attack crashed through the morning mist on a frontage of thirty-two kilometres. The Germans were taken utterly by surprise. They had not reckoned on a mass assault by armour, and they were powerless to meet it. Bayonets were completely useless. It was a matter of chance whether machine-guns, hand-grenades or Minenwerfer did any damage. Artillery alone would have been at all successful, if it had been employed in the right way; as things were, gunners would have had a difficult, almost impossible task when the feeble light of early morning was dimmed still further by natural and artificial fog, when dust and smoke were thrown up by the creeping barrage, and a confusing multitude of targets hove into view at very close range. In fact no useful guns at all were available in the German infantry's combat zone. What could the German troops in their wretched shell holes possibly do when they saw the tanks rolling towards them? If they fired on the machines, or the enemy infantry coming up behind, they would be detected by the tanks and wiped out; if they held their fire some of them might escape being seen and destroyed by the tanks, but the enemy infantry would approach unscathed and take them prisoner. In the circumstances of the combat of 8 August 1918 the German infantry were defenceless in the face of certain destruction.

The movement of the British tanks was timed to ensure that they were crossing their own front line when the barrage opened. The creeping barrage dwelt for an initial three minutes on the foremost German trenches, then lifted by one hundred metres every two minutes. Later on the progression became slower, lifting every three minutes, and then only every four minutes. The tanks and infantry followed a short distance behind the curtain of shells. In addition to the creeping barrage the British directed a heavy fire against batteries, approach routes, villages, camps and combat positions. Within a short time all communications were destroyed, the telephone lines were torn up and the signal lamps were unusable – only wireless communication remained largely intact, but the transmissions were incapable of giving any clear picture of the situation in the forward combat zone. Messengers and runners simply did not come back. Only one thing was clear – that the enemy were dealing an almighty blow.

The Germans girded themselves for action as best they could; all intact guns and Minenwerfer discharged an 'annihilation fire' into the fog; but they were pathetically few in number and they were shooting blindly at ground from which in all probability the enemy had already departed. Where could the guns safely fire without endangering our own troops? Exactly how far had the enemy reached? In what direction should the reserves counter-attack? Should the machine-guns in the rear open fire without positive identification of the enemy? A growing uncertainty gripped officers and men alike among the German reserves and batteries.

We shall now consider the course of the British attack in sequence from north to south:

On III Corps' sector 12th and 18th Divisions reached their objectives between 0730 and 0800.

In one of the episodes, 18th Division's single tank company attacked along the Corbie–Braye road, destroyed the greater part of 2nd and 3rd Battalions, 123rd Infantry Regiment, and had a clear run to the German batteries which were standing defenceless in the fog in the woods of Tailles and Gressaire. However the British officers obeyed their instructions to the letter – not to go beyond their closely defined objectives, and to observe the stipulated pause in the attack. The tanks remained motionless until the fog lifted, and the German batteries (which were still devoid of infantry support) were now able to hold their ground.

Accompanied by two companies of tanks, British 58th Division reached its objective long before the German infantry reserves could come up to protect their batteries. Here again the intermission in the British attack turned to the advantage of the Germans, who by 1115 had occupied and consolidated the important position of the 'canal hill' north of Chipilly, undisturbed by the enemy. When the fog cleared, after 0945, the Germans were able to direct a flanking fire against the Australians who were advancing south of the Somme, which disrupted their progress considerably.

Along the whole frontage of the British corps it was impossible to get the attack moving again after the pause; the second objective was never reached, and the single favourable opportunity for pressing on with the offensive had flitted by. A few British units got as far as the foremost German batteries, but the artillery was able to hold its positions.

We turn to the Australian Corps, where 48 of the available 144 tanks were assigned to the four brigades of the first line of infantry (Australian 3rd and 2nd Divisions) on a frontage of six kilometres. This wave was under orders to penetrate three kilometres and reach its first objective at 0720. A two-hour pause was then supposed to ensue, during which the second line of infantry, comprising Australian 4th and 5th Divisions with 96 tanks, moved forward through the first line. Despite the very scanty complement of tanks the first enemy wave was able to capture the German positions

TANK BATTLE AT AMIENS, 8 AUGUST 1918

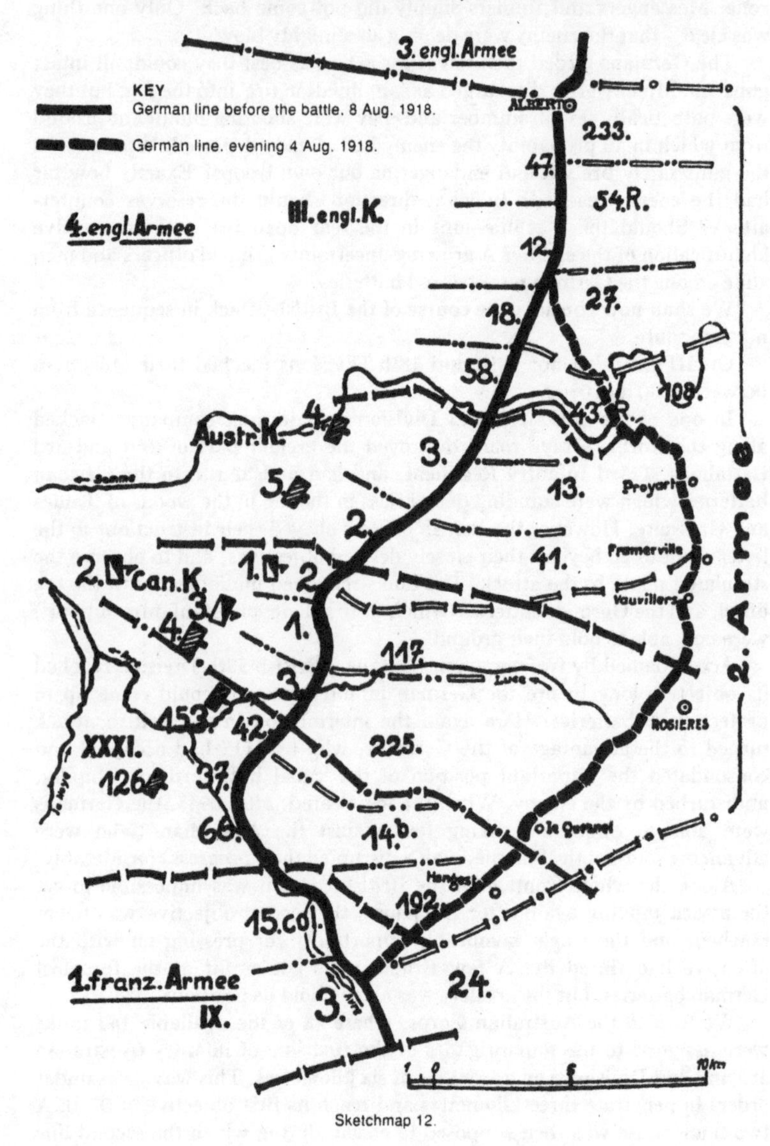

Sketchmap 12.

along the le Hamel–Cérisy road to the south by 0720, according to plan. It seems that the tanks now went counter to orders and exploited the opportunity on their own initiative – at least the fog was still lingering when they pressed on beyond the first phase line and proceeded to capture the batteries south-west of Cérisy. Not long afterwards an action broke out in 202nd Reserve Jäger Regiment's positions south-east of Cérisy; the fighting went on until the gallant defenders, having destroyed a number of tanks, finally withdrew north-east over the Somme. The tanks now overran 13th Jäger Division's combat and reserve battalions; as early as 0630 the severely wounded commander of its 13th Jäger Regiment was taken prisoner; a battery had already been lost at 0620, and the officers of 15th Jäger Regiment's were wounded and captured. The Australians reached their objectives at 0720, according to plan. There was absolutely nobody to defend the mist-covered ground which stretched between them and 13th Jäger Division's batteries which still had ten light and eight heavy guns serviceable. The two-hour pause now supervened, but the Germans failed to make any use of it and bring up reserves to protect the artillery. The Australians resumed the attack, and took the whole of the artillery between 0920 and 1000. Towards 1030 German reserves tried to hold the ravine south of Morcourt, but already by 1100 they had been enveloped by the tanks and their position was hopeless; half an hour later they were wiped out by the combined fire of aircraft, machine-guns and tanks, and the Australians reached their second objective on schedule. Only a few troops were available to the Germans to occupy Hill 84 west of Proyart, and after Australian infantry had paved the way the defenders came under attack at 1230 by aircraft and tanks working in concert. With this gain the enemy had attained their objectives for the day and brought their advance to an end. Quite possibly their aggressive spirit had been tempered by the effective flanking fire which, as we have seen, came from the intact German batteries north of the Somme.

To the south of Australian 3rd Division, the attack by Australian 2nd Division had likewise developed as was laid down, the assaulting troops penetrating the advanced batteries between 0700 and 0730. However the pause in the attack, together with the lifting of the fog, permitted the rearmost German batteries at Bayonvillers to inflict some local losses on the tanks, when they reappeared on the scene, and the gunners were able to hold out until 0950. The German infantry, or rather what was left of them, remained along the axis of the Villers-Bretonneux–Harbonnières–Lihons road until the Australians and Canadians resumed the attack, and they then fell back to the east of Harbonnières. The six German batteries around Marcelcave lay within the first enemy objective, and they were lost early on; the batteries lying outside the phase line at Pierret wood were able to keep up their fire for two hours more. At 0920 Australian 5th Division with its fresh tanks passed through 2nd Division on its way to the next objective. All that the Germans could set against them by that time were seven companies of riflemen and three companies of machine-guns, with a weak battalion behind. There was no artillery left. The German reserves essayed a counter-attack against Bayonvillers, but they were halted by enemy tanks and aircraft at the 'Roman gorge' 2+ kilometres to the north-east. 'The battalion broke up under the impossibility of offering any defence against tanks. It was literally smashed to pieces.' (*Schlachten des Weltkrieges*, XXXVI, 124.) At this juncture the armoured cars of 17th Tank Battalion appeared on the Roman Road. They were moving so fast that the German gunners were unable to take a bead on them, and they played havoc among the columns of German vehicles seeking to escape. The British aircraft dropped smoke bombs on the Roman gorge, which blinded the defenders further to the east and helped the British armour to approach unobserved. The enemy took Harbonnières, and reached their third objective at noon.

The Canadian Corps, which also had a reputation for aggressiveness, was on the attack to the south of the Australians. Canadian 2nd Division's assault was spearheaded by a tank battalion, and the impact fell mainly on 148th Jäger Regiment of 141st Jäger Division; the corresponding push of Canadian 1st Division was likewise led by a battalion of tanks, and it hit 117th Jäger Division. This was an excellent formation which was fully up to combat strength, and it had a fine leader in Major-General Hoefer. All the same, two of the infantry regimental commanders of the division were taken prisoner, and the third died a hero's death. The Canadians had slipped through the German barrage, which was not particularly powerful, and taken their first objective, and with it most of the German batteries. Every attempt at resistance was broken by the way the Canadians mounted a turning movement to the north, and only when the Canadians halted, after reaching their third objective, were the Germans able to form a new front west of Rozières.

Canadian 3rd Division struck at 225th Jäger Division and very rapidly broke into the valley of the Luce, though further south the 'black wood', which extended behind the main road from Domart-sur-la-Luce to Mézières, was impenetrable to tanks, and the Germans were able to stand there until about 0830. In spite of this holdup the attack on the second objective opened up more or less on time and reached its goal at 1030. The Germans had only fifteen guns with which to oppose the advance, and they were unable to bring up any infantry reinforcements. Supported by fresh tanks, Canadian 4th Division and British 3rd Cavalry Division continued the push in the direction of Beaucourt. The time was shortly after 1100.

Between 1100 and 1200 the two corps, Australian and Canadian, managed to break through the German forward battle zone, and take all the German artillery with the exception of a few guns. The only forces that the Germans could throw against them were a few battalions which had been resting in the rear; while they were still on their way they were badly mauled by aircraft and long-range artillery fire. So feeble had the German resistance become that in places the enemy were able to proceed in column of march. Now it was just a question of completing the breakthrough.

The British high command believed that the Cavalry Corps was a particularly suitable instrument for this purpose, and it had tried to augment its striking power by putting under its command the two newest and fastest tank battalions, which fielded 96 Whippets, as we have seen. These, inevitably, were parcelled out among the various formations.

The Cavalry Corps formed two lines. British 1st Cavalry Division was ordered to overtake the infantry north of the Luce as rapidly as possible, while British 3rd Cavalry Division did the same to the south of the stream; they were to reach the third objective, wait there until the infantry arrived, and then push on to the Chaulnes–Roye road, with the British 2nd Cavalry Division following in the second line. The Whippet battalions drove in front of the first-line divisions, so as to cover the horse soldiers and open lanes through the wire. By 1015 the first-line divisions had reached a line between Ignaucourt and Marcelcave, and they now deployed to fulfil their missions. Sixteen tanks were allocated to each brigade of three cavalry regiments and a mounted battery. They then advanced in the following order:

1st Cavalry Division – 1 Cavalry Brigade against Harbonnières; 9 Cavalry Brigade by way of Guillaucourt against Rosières-en-Santerre; 2 Cavalry Brigade against Caix;

3rd Cavalry Division – 7 Cavalry Brigade through Cayeux against Caix; 6 Cavalry Brigade against le Quesnel; Canadian Cavalry Brigade against Beaucourt.

First Cavalry Brigade penetrated the farthest, coming to a halt in front of Framerville and Vauvillers. The others never got as far as their third objective, which was where their main task, the breakthrough to the railway between Chaulnes and Roye, was supposed to begin. It is not too much to say that the cavalry could not have progressed as far as they did without the protection of the tanks. Where a mounted advance was attempted in larger formations, it invariably collapsed with heavy casualties in a matter of minutes, as witness the action of 6 Cavalry Brigade south-east of Cayeux or the Canadian Cavalry Brigade's at Beaucourt, even though the Germans had no coherent line of defence between Caix and Beaucourt. Just 2½ companies of German engineers, standing in Caix and to the south-west, were sufficient to bring 3rd Cavalry Division's advance to a halt, and they gave way only when the tanks opened their attack and drove them back to the north of Beaufort. No more than a few troops of cavalry were able to move here and there over the battlefield. The second line of cavalry was not sent into action.

Towards noon 17th Tank Battalion with its twelve armoured cars advanced through the lines of infantry and cavalry, which were still pinned to the ground, and penetrated through and beyond the villages of Framerville and Proyart. The tanks wrought dire confusion in the rear of the German front lines, inflicted heavy losses on columns and reserves, and held on in the neighbourhood of the two villages for several hours without suffering any losses – but also without any attempt on the part of the other British forces to follow them beyond their third objective. It was not until about 1830 that the German reserves entered Proyart and Framerville, finally closing a gap several kilometres wide. For six hours, until the late evening, it had stood open to a British breakthrough; there had been hardly any German defenders in the way, and effectively no artillery at all. But the British were unwilling to depart from the battle plan because it had been laid down by generals who were protagonists of combat by infantry, artillery and even cavalry – in other words by men who did not know how to make use of the most powerful punch available to them. British aircraft successfully attacked and pinned down the German reserves, tanks were at hand in considerable numbers, and no German resistance to speak of had been encountered. All of this was of no avail – the British did nothing.

The French XXXI Corps was attacking south of the Canadian Corps. Contrary to the procedure of British neighbours, the French put in a 45-minute artillery barrage before they attacked. The first assault was to be delivered by three infantry divisions without the support of armour, but then the French 153rd Divsion was to come storming through accompanied by three battalions of tanks. The success of the initial break-in was due to the collapse of the Germans who faced the Canadians further north. Very shortly afterwards 153rd Division went into action with its tanks and made some tangible progress. However at times the French attack hung well behind that of the Canadian right wing, where the flanking fire of the

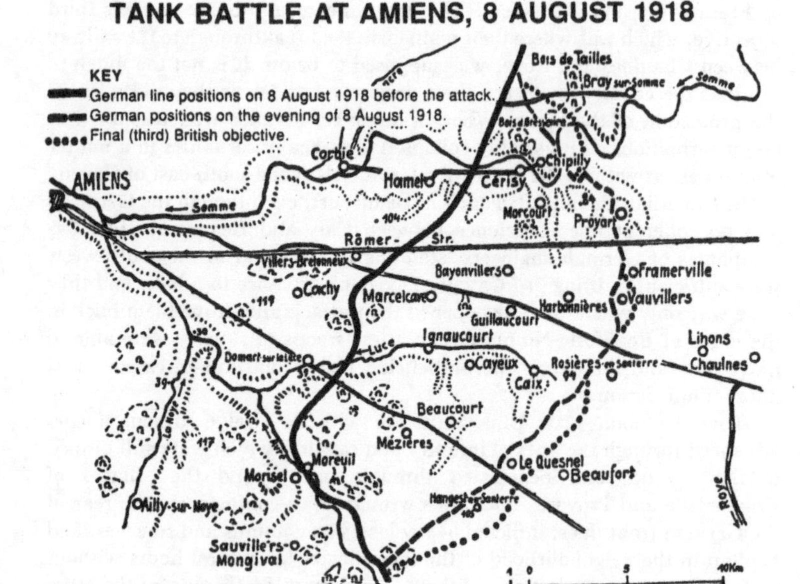

Sketchmap 13.

German artillery was consequently able to inflict heavy losses on the tanks supporting the Canadian attack.

The battle burnt itself out when the summer night descended on the field between 2200 and 2300, and '. . . by then the German Army had experienced its greatest reverse since the beginning of the war'. (*Schlachten des Weltkrieges*, XXXVI, 196.) Eight divisions had been almost completely smashed, and eight more had been hard hit; in a matter of hours the Germans had lost 700 officers and 27,000 men (including 16,000 prisoners) and more than 400 guns. The enemy had effected a break-in on a breadth of 32 kilometres and a depth of 12 kilometres; not until evening were the strenuous efforts of the Germans able to establish a new line, and they owed it to the inactivity of the enemy. It is true that our enemies had been unable to accomplish a worthwhile operational breakthrough, and it is equally true that there was no longer any immediate danger of a complete collapse of the Western Front. However the effect of this mighty blow on the German high command was, inevitably, extraordinarily great.

Even now old-timers like us re-live that feeling of impending doom which overtook us on that day in August. General Ludendorff had been the soul of German endurance for two years, and we can imagine how he felt when even he was forced to conlude that the war must be brought to an end, and that even his own titanic strength could not avert our fate. In reality there was no other way out. The Battle of Soissons led to the disbandment of ten divisions; a few weeks later the events of 8 August cost us the same number. The German Army's power of resistance faced an inevitable decline, while fresh forces streamed ceaselessly to the armies of the enemy alliance. One million Americans, and incalculable numbers of tanks and aircraft were ready for the Alliance as early as the autumn of 1918. There was absolutely no prospect that the military situation would be any better for us in 1919. Ten days after the Battle of Amiens an Imperial council in Spa decided to open peace negotiations at a suitable opportunity. Meanwhile the war was to be conducted on the defensive.[12]

Before we follow the operational and political consequences of 8 August any farther, we should pause to consider the tactics of the two parties, and those of the enemy in particular.

As we have seen, the Germans had planned their defence to meet a conventional artillery and infantry attack, but they had done virtually nothing for anti-tank defence. There was no attempt to establish the main line of defence in terrain that was impassable or even difficult for tanks, and no attempt was made to site the artilllery – our only effective anti-tank weapon – for direct fire. There were no other means of defence at all. The experiences of 18 July and 8 August teach us that any serious defence against tanks must be established behind tank-proof obstacles if we are not to expose our infantry and artillery to annihilation. Without such protection it is of no real use in itself to equip the infantry divisions with a full array of anti-tank weapons, since the effect of these devices depends on so many chance considerations. In the future, combat in the open field will be unthinkable unless we can match the enemy in terms of tanks.

As for the attack, the battle of 8 August represents the third total success for the Cambrai recipe. We must applaud the decision to renounce artillery preparation in favour of a surprise attack, the painstaking measures of concealment, and the teamwork among the assaulting troops. The effectiveness of the attack rested on the tank, and the chosen ground was obstacle-free. There were 500 tanks in the push, which was the same as at Soissons and only 100 more than at Cambrai;[13] it is therefore wrong to speak about an 'undreamt of increase' or 'an unprecedented mass' of tanks. (*Schlachten des Weltkrieges*, XXXVI,221,222.) An 'unprecedented increase' in tanks is something which lies in the future.

In some respects the British could have done better with their tanks than they did. It would have been perfectly easy for the British commander-in-chief to have had a far greater number of machines available for 8 August, if only he had recognized more clearly the battle-winning potential of tanks, and exerted greater pressure on the manufacturers in Britain. The breadth of the tank attack was broader than at Cambrai, it lacked depth, and it was tied in too closely with the infantry and the Cavalry Corps, especially in the second line, and it therefore lacked the speed and independence which were needed to exploit the initial success.

The artillery tactics matched the objectives of the offensive and the way the tanks were intended to fight. On the other hand the objectives themselves should have been selected better than they were, after the previous experiences of the British. We may cite the two-hour halt not long after the attack began, the closeness of the first objective, and the fixing of the final objective for the tanks and the infantry divisions immediately before the positions of the German reserve divisions. Only these miscalculations on the part of the British enabled the feeble German artillery to recover some of its effectiveness, or the defenders to establish a weak new line on the evening of the battle. It would have been a different story if the British had liberated their tanks from the conventional arms, which were inherently slow and vulnerable to machine-gun fire, and if they had struck simultaneously into the full depth of the German defensive system, the details of which were perfectly well known to them. Then it would have been a moral certainty that the British would have destroyed the defenders in short order and broken through their entire front.

At Amiens the British had available:

(a) for action against the German reserve divisions and command centres – the Whippets and armoured cars; as well as a large number of aircraft;

(b) for action against the German artillery and infantry – the two lines of heavy tanks, of which the second was comparatively weak, since it only had to destroy the defenceless German infantry; this line was the only element which needed to be tied to the pace of the British infantry in the early stages, and otherwise there was nothing to prevent the armour taking full advantage of its speed.

By noon on 8 August, despite their generally slow procedures, the British were standing at the equivalent of an open door. And yet again the cavalry showed themselves quite unfitted to appear on the modern battlefield, just as at Cambrai and Soissons. Although von Schlieffen recognized this fact in 1909 and presented it in his article in unchallengeable terms (see p. 31), we are still reading of people who try to argue the opposite, and call for the reintroduction of cavalry at the army level – in fact for heaping up human beings and animals which have little inherent combat value. They will achieve nothing in a future war, now that we have to reckon with the multiplication of machine-guns, the buildup of tanks and aircraft, and the possible employment of chemical weapons. Cavalry now have only an insignificant advantage in speed over infantry, especially when we make the comparison with motorized forces; in recent years the development of off-road vehicles, and especially those with tracks, has virtually equalled, and probably even surpassed, the much-vaunted superiority of the mounted arm in cross-country mobility. In all other respects cavalry are at an outright disadvantage. Until 8 August Sir Douglas Haig had sedulously spared his 27 regiments of cavalry. No doubt the horsemen had plenty of élan and offensive spirit, but these qualities would have shown to much better advantage if the cavalry had gone into action in the form of tank squadrons. As it was, they were thrown away in fruitless attacks. 'In a matter of minutes the mounted attack collapsed in the face of our fire, which was extraordinarily violent, especially from our heavy and light machine-guns. I shall never forget the sight – how the cavalry pushed forward, and was converted in the next instant into a mass of horses, weltering in their blood, hobbling on shattered limbs, or galloping riderless through our lines of infantry.' (*Schlachten des Weltkrieges*, XXXVI,186.)[14]

Von Schlieffen's verdict is equally valid in other respects. Mounted cavalry represent a large and vulnerable target, and they are helpless to a frightening degree, even whey they are accompanied by tanks. Things can only get worse. There is every prospect that tanks will continue to develop at the present impressive pace, whereas the performance of the horse cannot be improved to any significant degree; in other words, the gap between the two arms will expand rather than diminish, and any attempt to bind the unequal partners together can only act to the disadvantge of the tanks, and consequently of the whole.

We end with a review of the continuation of the battle between 9 and 11 August. There was nothing new in what was achieved or how it was done. Excluding five tanks which broke down, the armour lost: 8 August: 100 machines out of the 415 in action; 9 August: 39 out of 145; 10 August: 30 out of 67; 11 August: an unknown number out of 38.

Those 100 tanks that were lost on 8 August had been knocked out by the German artillery, which took advantage of the respite granted by our enemies; however the German losses in guns came to 400, which was not a particularly advantageous trade-off.

The Battle of Amiens caused no alterations of tactics on the part of either the attackers or the defenders. Changes were probably ruled out in any case, given the speed of events over the following weeks, the falling-away of Germany's allies, and the decline in the combat-effectiveness of the German Army itself. However we must be careful not to regard the events of 1918, and especially the battle of August, as some kind of culmination. On the contrary; those events were the beginning, not the conclusion of a total revolution in tactics and therefore of operational potentials. This revolution was the result of the appearance of new weapons in ever greater numbers. Why, then, did the weapons in question not bring about a total collapse of the defence every time they went into action? The reason is that they were underestimated at the time of the war, and by the Allies no less than the Germans. This misappreciation led in turn to the misuse of those weapons in the field.

3. THE END OF THE WAR. THE WAR IN THE AIR. TANK WARFARE. CHEMICAL WARFARE. U-BOAT WARFARE.

French successes on 18 August 1918 and British on 2 September were both scored through a heavy commitment of tanks, and they led to the withdrawal of the German front to the Hindenburg Line, from where the German offensive had embarked with such high hopes in the spring. On 12 September the Allies took the Saint-Mihiel salient between the Meuse and the Moselle, assisted by 232 French tanks which attacked from the south, exploiting the most favourable ground. However the French tanks came to a halt in the afternoon and were immobilized for the next twenty-four hours, simply because the American military police would not allow their convoys of fuel to pass through.

On 15 September the German high command wrote to the Kaiser: 'No doubt can remain that the enemy will continue their offensive through the autumn. They have the necessary means, thanks to the stream of American forces and the mass employment of tanks. On our side we will keep up the fight, not to hold ground as such, but with the object of letting the enemy wear themselves out while we maintain the battleworthiness of our own army.' (Schwertfeger, *Das Weltkriegsende*, 100.) This strategy fitted the circumstances of the time, though it does not seem to have been followed through with sufficient vigour. In any event our continuing combat losses and the decline in our strength led to the disbandment of further divisions, the reduction in the establishment of the battalion from four companies to three, and in some cases the diminution of the battalions in the regiment from three to two.

On 15 September 1918 the Austrians published a peace note which brought home just how serious the situation was. On the same day the Bulgarian front collapsed in Macedonia, and the Turkish front in Palestine followed suit on 18 September. On 25 September the Bulgarians sued for peace. 28 September brought a conference between Field Marshal Hindenburg and General Ludendorff, in which they decided that the war must be brought to an end through a request for an immediate armistice. The Kaiser gave his assent the next day, and a further consequence was the re-casting of the government on the parliamentary model.

On 30 September General Ludendorff declared at a conference that 'The conduct of the war on the Western Front has been reduced to a game of chance, due chiefly to the effect of the tanks; the high command no longer has a firm basis for its calculations.' (Schwertfeger, 128.)

Berlin was the scene of the fateful gathering of the party leaders on 2 October, when a representative of the high command reported on the situation at the front. He urged the necessity of an armistice, citing the 'action of the tanks, which we are unable to counter, and the state of our replacements'. (Schwertfeger, 134.) On 3 October the German Government addressed a request for an armistice to the President of the United States.

This deserves to be underlined. When the representative of the high command asked for an immediate armistice he did so on two grounds, of which the first was the enemy superiority in tanks. It is fair to assume that the deputy was as well informed on the views of the front-line soldiers as he was about the thinking of our military leadership. This was a grave and tragic hour, and objective calculations alone could have induced the high command to make such a request. As for the reasons presented at the meeting, these too must have been the product of weighty and conscientious examination.

The war was continued in the form of a costly defensive battle until the armistice came into effect on 11 November. On 26 September the Americans attacked between the Argonne and the Meuse with 411 tanks; the French Fourth Army opened a simultaneous push with the support of 654 machines. The British joined in with an offensive at Cambrai on 27 September, followed by the Belgians in Flanders on 28 September.

The details of the American attack are instructive. Co-operation with the armour did not work particularly well, and the reasons are the unsuitable terrain chosen for the advance, and the frequent counter-orders which were issued from the second day of the attack onwards. On many occasions the infantry failed to exploit the armoured successes, with the result that a great number of tanks fell into the hands of the Germans. The conclusion must be that even completely fresh, unused and fully battle-worthy infantrymen, like the Americans, showed that they had no inherent striking power in the face of machine-guns, and were often incapable of following the armour, slow-moving though it was in those times. To a still greater degree than with the British and French attacks, the battle on the second day degenerated into costly and unproductive local actions. The French Fourth Army went about things differently. They opened their attack across the shell-cratered landscape of the old battlefields, but they did not commit their tanks until they had taken the ground and made it practicable. To this work they assigned 2,800 labourers, who completed it on 28 September, along with clearing lanes through the anti-tank obstacles, mines and trenches. This army lost only two tanks to mines.

In the fighting on 27 and 28 September it became clear that the French infantry no longer had the strength to utilize the success of the tanks. We frequently encounter phrases such as the following: 'The tanks put the defenders to flight, but the infantry failed to reach their objective.' (Dutil.) This shortcoming became the more evident the more the battle degenerated into scrappy combats. Conversely, the better the attack was directed, and the greater the number of tanks that went into action, the more rewarding were the results. On 29 September there was a temporary falling-off in the combat readiness of the machines, and only a few units took part in the actions on the 30th. By 1 October altogether 180 tanks were once more ready for service, and the French had taken a total of 12,000 prisoners and 300 guns. The French resumed the offensive on the 3rd. It went well, though by 8 October, when Fourth Army's tank forces were exhausted, they had lost 40 per cent of their officers, 33 per cent of their men, and 39 per cent of their tanks. Of the 184 tanks which fell out of action, 56 had been disabled by artillery fire, and two through mines, while the rest had broken down; 167 were repaired and soon back in service, 17 were lost for good and two were missing.

Tank units were used virtually everywhere in the fighting that took place during the pursuit in October, though many of them had to drop out through mechanical wear towards the end of the month. The main enemy turned out to be field guns employed in the anti-tank role, and sometimes also Minenwerfer firing at low trajectory. By way of contrast the losses to mines were few, evidently because the minefields were inadequately camouflaged, but perhaps also because the enemy might have been informed as to their whereabouts.

The salient feature of all the October battles was the way the tanks were flung into combat without any planning. All the lessons of 18 July and 8 August seem to have been cast aside, and there was not a single occasion on which the huge quantities of tanks now available (in the order of 4,500 at the very least) were directed at a common objective, simultaneously, and with due co-ordination. Indeed, the choice of ground for attack often seems to have been made from political rather than military considerations. There was no real need for haste and disorder of this kind – after all, the Allies knew the prospects for the Germans only too well!

By 1 October the French had 2,653 tanks, and the monthly production had attained a rate of 620 machines.

The Armistice brought an end to hostilities on 11 November 1918.

We must now ask how well the various categories of weapons were performing by the end of the war. We will then proceed to examine the lessons which influenced post-war developments.

Until it ended, the fighting on the Western Front bore essentially the character of positional warfare, even if very few properly engineered defences were constructed in the closing weeks. This was a type of combat in which the machine-gun became the dominant weapon, and made life difficult or impossible for unprotected soldiers and horses. The positions ultimately evolved into wired, dug-in machine-gun nests which were secured by outposts and communication trenches, and which relied for close-range defence on the hand-grenade as well as the machine-guns themselves. To the rear extended a deep array of covered batteries and troops on standby.

Unsupported infantry were quite unable to tackle a defence of this kind. The machine-guns offered such small and insignificant targets that the attacker required a huge outlay in artillery and ammunition to silence them – and if only a few machine-guns survived they could usually shatter an attack by considerably greater forces. In the best cases the cost of an infantry attack was entirely disproportionate to the gains, not least because the facility of motor transport made it possible for the defenders to bring up their reserves in good time; conversely, instead of the intended breakthrough and operational exploitation, the attacker had to be content with making a salient into the enemy line, with all its attendant tactical disadvantages.

It follows that if infantrymen are reduced essentially to becoming machine-gunners, the negative strength of their weapons makes them primarily suitable for the defensive role in the post-war world. The offensive power of infantry extends no further than the range of their machine-guns and the other integral weapons, and even then only after everything that lies beyond their limited objectives had been suppressed by other weapons, and in particular the artillery. If the artillery in question fails in its task of total suppression – silencing the greater part of the machine-guns, crippling the batteries, and in general conquering ground by fire – one's own infantry will be unable to occupy and hold it.

The conquest of an objective by artillery demands a considerable expenditure of ammunition by a great number of batteries. Preparing an artillery attack therefore becomes a time-consuming and highly-visible business, which detracts from the chances of achieving surprise. If short artillery bomardments are desirable in principle, their actual effect is open to question. Long artillery bombardments reduce the area of the attack to a landscape of shell holes, impede the movement of vehicles from the rear, and the exploitation of any initial success. The depth of the artillery attack depends not so much on the actual range of the guns as the facilities for observation, and to carry it through to success depends on an accurate knowledge of the enemy deployments, otherwise we will be unable to engage the targets directly and will use up more ammunition than we can replace. The artillery, like the infantry, has to limit its attack to one objective at a time, because it must change position afterwards. This pause turns to the advantage of the defence. When the attack is resumed – usually in not such a systematic way as in the first bombardment – it does not fall on empty space, but upon a new defensive line where the enemy deployment is usually unknown, and which is consequently a hard nut for the infantry to crack.

While the offensive striking power of artillery proved to be far more formidable than that of the infantry, it remained too slow-acting, too wasteful and too subject to chance to be able to secure a rapid breakthrough.

The cavalry had been the third main arm in 1914, and yet by 1918 its useful work had been confined to carrying messages and executing short-range reconnaissances under the command structure of the infantry divisions. Otherwise the cavalrymen were reduced to mounted infantry, and they should be assessed accordingly.

It was the other way around with the air forces, which were used mainly for reconnaissance at the beginning of the war, but became an extremely important weapon as the conflict wore on. Aircraft in the reconnaissance and artillery observation roles were a nuisance to the enemy ground troops, because they forced them to take various measures of concealment and make use of the hours of darkness, but it was the ground-attack aircraft that became the immediate threat. The Germans suffered from the attentions of the enemy aircraft on the Somme and at Ypres, and in the course of 1918 the superiority of the Allies in the air became more tangible still. While enemy air raids against the German homeland were rare and not particularly effective, aircraft intervened to significant purpose in the ground battle, as at Amiens on 8 August 1918. They created disorder in the German rearmost communications, they hindered the movement of reserves, they took German batteries under actual attack, they laid smoke-screens in front of occupied ground, and they reported the progress of the attack. All of this was of material influence on the course of the ground fighting, especially when the aircraft were acting in co-ordination with tanks. Aircraft became an offensive weapon of the first order, distinguished by their great speed, range and effect on target. If their initial development experienced a check when hostilities came to an end in 1918, they had already shown their potential clearly enough to those who were on the receiving end.

To act to decisive effect, the air forces still need a partner on the ground that is able to overcome the defensive strength of modern weapons speedily enough to expand the break-in to a full breakthrough, exploiting both the initial success of the offensive and the work of the aircraft. The conventional ground forces too stand in need of a partner of this kind – and in fact in future wars they will have little offensive capacity without it. This partner, this new weapon, is represented by the tank. We have already described in some detail the influence which armour exercised on the conduct of war since it first appeared in September 1916. We have not touched on the reasons why the Germans virtually renounced the idea of creating tank forces of their own, but the consequences of this failure are only too evident, and they were made still worse by the lack of any suitable anti-tank weapons, or even an intelligent use of the available artillery for this purpose.

The importance that the enemy attached to tanks may be deduced by the construction programme for 1919. The Allies intended to increase their armoured establishments as follows: Britain from 2,000 machines to 7,000;

19. Italian Fiats Unfaldo, 1933.

20. An artist's impression of an idea for an air-portable tank as proposed by the American inventor Walter Christie.

21. In May 1940 the Germans used paratroops in advance of major armoured thrusts in much the way Guderian contemplated in *Achtung–Panzer!* The photograph shows the drop on Rotterdam.

22. A British Vickers Guy armoured car with optional track on the rear wheels.

23. French Panhard-Kegressé-Hinstin M29.

24. French 'Dragons Portés' on armoured cross-country vehicles. (Panhards-Kegressés-Hinstins 16 CV)

25. The Austrian Uchtrad-Stenr armoured reconnaissance vehicle.

26. German armoured troop-carrier as permitted by the Versailles Treaty, 1922.

27. Uttrappe armoured reconnaissance vehicle as permitted under the Versailles Treaty, 1931.

28. German light armoured reconnaissance vehicle, 1935.

29. German armoured reconnaissance vehicles. The second vehicle carries a radio antenna.

30. German armoured reconnaissance vehicles moving through woodland.

31. German machine-gun tank (Panzer I) 1934.

32. In the forest.

33. In the snow.

34. Going cross-country.

35. Crossing a stream.

36. A Panzer II.

37. A German 3.7cm gun in action.

38. German Schützen (mechanized infantry) on motorcycles.

39. German Schützen in trucks.

40. A portrait of Heinz Guderian.

France from 2,653 to between 8,000 and 10,000; The United States to 10,000.

By way of contrast the Germans intended to build up only from 45 to 800.

While the British concentrated on building heavy and medium tanks, the French and Americans up to 1918 put the emphasis on the light Renault. General Estienne, however, was of the opinion that the French would be faced with the problem of overcoming formidable German defences in the following year, and as early as February 1918 he urged the building of heavy tanks: 'The decisive attack will follow in the wake of the heavy tanks, which will smash a way through the entire zone of obstacles, and not just for the infantry, but the horse-drawn artillery and the other tanks as well. The infantrymen will be following immediately behind the light tanks, which are their loyal and inseparable companions, and they will have every confidence that success on this first day will give the offensive an added impetus, instead of exhausting it.' (Dutil, 26.) Estienne had in mind a continuous offensive which would incorporate surprise at the strategic level, and was to be sustained by a rapid movement of reserves and supplies. It seems that the British minister Winston Churchill was thinking along the same lines when, in July 1918, he told the Imperial General Staff that full reliance could be placed on the construction figures predicted for 1919, and that it should therefore lose no time in working out the most suitable tactics for the offensive. The Armistice supervened before our enemies could put their designs into practical effect, but the tank augmentations intended for 1919 are proof enough of their intentions. It is clear that the tank, as the embodiment of the ground attack, had risen to rank alongside the air forces in their offensive role.

Chemical substances were the third of the new weapons to arise in the World War. They suited the defenders to just about the same degree as the attackers, and so they cannot be counted as an exclusive asset of the offensive. For offensive purposes, short-term chemical weapons are used on ground over which our own troops will be attacking. Conversely it suits the defenders to employ persistent agents to contaminate terrain over a length of time – this is particularly useful in the case of a retreat, for it may help our forces to disengage.[16] Motorized troops are in fact the only forces that are capable of crossing contaminated ground at speed.

The U-boat was the fourth device to take on an unsuspected significance. The Germans had stolen a march on their enemies in technical development, and the war might have taken a different course if the German government had had the nerve to make timely and unrestricted use of the weapon.

Sooner or later every new weapon conjures up countermeasures. Offensive air power has already been met by anti-aircraft units utilizing artillery pieces, machine-guns, searchlights and nets, camouflage and blackouts, and by fighter aircraft which can meet the enemy in their own element.

Chemical weapons can be rendered ineffective by masks and protective clothing, or by chemical neutralizers.

The Allied war against U-boats was conducted by nets and destroyers, aircraft and depth-charges, the convoy system, but chiefly and most effectively by propaganda and diplomatic pressure. Germany allowed itself to be intimidated.[17]

As we have already noted, the least of all was done for anti-tank defence. Neither suitable guns nor machine-guns were deployed by the end of hostilities, and the German 13mm anti-tank rifle was largely ineffective. Only the engineers went to the trouble of putting up any kind of defence, when they set out obstacles and mines. In contrast the artillery failed to adopt the new tactics that were needed to render speedy and efficacious help to the infantrymen against the tank, which was a particularly dangerous enemy of theirs. This was particularly relevant during the war, when artillery was the only useful weapon against armour. Things have admittedly changed a great deal since then.

For centuries now the Germans have regarded the infantry as the principal arm of service, and in the World War they assumed that infantrymen would be equal to any task that was set before them, no matter how difficult, and not excluding anti-tank defence. This was demanding altogether too much.

To sum up, two of the new weapons that emerged in the World War, tanks and aircraft, predominantly augmented the power of the defence, while chemical devices and U-boats served the offensive and defensive modes to an equal degree.

At the time of the war engine-powered offensive weapons were in their infancy, and even now they stand only at the beginning of their development. As early as 1918, however, they worked to decisive effect, and our victorious enemies concluded that they must henceforward deny them to the Germans.

Notes

1. The best modern account of the technical development of British tanks in the First World War is David Fletcher, *Landships* (HMSO), 1983, passim.

2. On this truly remarkable collection of personalities the best contemporary source is a privately printed booklet *Tank Corps H.Q.* which is in the library of the Tank Museum at Bovington. All the personalities are referred to by pseudonyms but the Tank Museum's library staff have managed to provide a key.

3. In fact Fuller claims that at this stage he had never seen a copy of Swinton's paper of February 1916, though he admired it when, early in 1918, he finally did see it. Fuller, *Memoirs of an Unconventional Soldier* (Nicholson and Watson), 1936, p. 169.

4. This conclusion is a vitally important one for German armour doctrine as Guderian preached it. Guderian had an acute sense of the limitations of tanks and of their need for co-operation with infantry and artillery. In Great Britain, however, in the late twenties and thirties, some of the leading lights of the Royal Tank Corps became impatient with the problems of co-operating with truck-borne infantry and towed artillery, which had different mobility characteristics from theirs. The habit of mind which developed in the Royal Tank Corps avant garde is well illustrated in remarks made by Colonel Eric Offord during an interview with the American historian Harold Winton. 'We didn't want an all tank army but what could we do? The infantry were in buses (i.e., trucks), they couldn't come with us. The artillery were . . . obstructive. They never put the rounds where you needed them; and when you called it always came too late.' Winton, *To Change An Army* (Brassey's) 1988, p. 110, note 23. The RTC developed an increasing tendency to operate tanks on their own, with scant regard for infantry and artillery co-operation. This tendency was later to have disastrous consequences on some occasions in the Western Desert.

5. There is no documented, scholarly account of Cambrai. Guderian's version appears to be based on published German accounts and on Fuller's *Mem-*

oirs. Liddell Hart's short account in *The Tanks*, Vol. I (Cassell), 1959, pp. 12853, appears carefully researched and is very readable, but naturally very partisan on the Tank Corps' behalf and not documented at all. A longer account – Bryan Cooper, *The Ironclads of Cambrai* (Souvenir), 1967 is also undocumented.

6. On 30 April 1918 Haig issued an order reducing the establishment of the Tank Corps in France by one brigade and three battalions – about a fifth. In fact, however, the Tank Corps was well below its projected establishment and Haig's proposal does not seem to have involved the disbandment of existing units. Some Tank Corps personnel (three battalions) were to be used, temporarily, as Lewis gun detachments, in view of a shortage of infantry in France. For Haig's order and Fuller's predictably hostile reaction to it see Haig to War Office, 30 April 1918, Fuller Papers, B45, Tank Museum, Bovington. See also Fuller, *Memoirs*, pp. 26977.

7. Guderian here rightly identifies the great unsolved
problem of the German Army in 1918 as that of pursuing the enemy vigorously once he had been thrown out of his defensive positions. In British military parlance this was called 'exploitation'. By employing the technique (developed by their artillery expert, Colonel Bruchmüller) of relatively short but extremely intensive artillery bombardments – coupled with the infiltration tactics of the Stormtroops – the Germans were able to smash right through Allied defensive positions, reaching the gun line on several occasions in the spring and early summer of 1918. The inadequacy of the German arrangements for pursuit, however, allowed Allied armies to retreat in relatively good order and eventually to regain their cohesion. John Terraine, *To Win A War* (Sidgwick and Jackson), 1978, pp. 5974.

It does not suit Guderian's purpose to emphasise in this book the progress that had been made in infantry tactics in the German Army between 1914 and 1918. For a recent account of this see Bruce Gudmundsson, *Storm Troop Tactics* (Praeger), 1989, passim.

8. Guderian writes that the Germans never achieved *Durchbruch* (breakthrough) in 1918. He is in some danger here of confusing English language readers. The Germans did smash right through the defensive positions of, for example, the British Fifth Army on the Somme in March. The problem was that no Allied army actually disintegrated (as might have happened if the Germans had found some way of mounting more vigorous pursuits) and thus the Allies were able to maintain a more or less continuous, if sometimes very ragged, front. Guderian here uses the term *Durchbruch* (breakthrough) to mean irretrievable disruption of an entire front rather than merely penetration right through a particular defensive position. The latter he refers to as a break-in. Most British military authors would tend to use the term 'break-in' to describe a penetration part-way through a defensive position, 'breakthrough' to describe penetration all the way through and 'exploitation' to describe the process of pursuing an enemy with a view to making the penetration of his position decisive. A British military author would be more likely to say of the Germans in 1918 that they achieved breakthrough repeatedly but found no effective means of exploitation.

9. Guderian here identifies the Battle of Soissons as the critical turning-point of the Western Front campaign of 1918 and, by implication, of the war as a whole. This battle is almost completely unknown in Great Britain by the title Guderian gives it. Those interested in the military history of the First World War know it as the 'Second Battle of the Marne'. But considering its great significance it is remarkable how little is available on it in the English language. There is no modern, scholarly book devoted to it. *The Two Battles of the Marne* (Thornton Butterworth), 1927 contains brief accounts by Foch and Ludendorff.

10. It is not clear on what evidence Guderian believes that the German tank attack at Villers-Bretonneux speeded up British tank delivery, but it is true that cuts in the Tank Corps earlier decided on were cancelled not long after this battle. Liddell Hart, op. cit., p. 167.

11. Guderian probably used Fuller's account of Hamel. His opinion of this battle echoes Fuller's exactly. Fuller, *Memoirs*, pp. 289-90.

12. Here, discussing 8 August, and again when describing the events of 30 September 1918, Guderian effectively dismisses the 'Stab in the Back' myth which Hitler propagated in *Mein Kampf*. Hitler seems to deny the reality of the army's defeat and blame the unsatisfactory end to the war on the treachery of Jews and Socialists. (A. Hitler, *My Struggle* (Hurst and Blackett), 1933, pp. 91-2.) Guderian admits with refreshing frankness the conclusive defeat of the German Army on the Western Front. It is perhaps surprising that in a supposedly totalitarian state a senior officer could, in effect, contradict the published opinion of the head of state not only with impunity but without the intervention of a censor.

13. Guderian's figures are open to question here. The regimental history gives the Tank Corps' total strength at Amiens at 604 though this includes supply tanks. The fighting tanks included 324 heavy tanks and 96 Whippets – a total of 420, of which 415 saw action. The same work is less precise about the figure for Cambrai, merely saying that it was 'over 300'. Liddell Hart, *The Tanks* Vol. I (1959), pp. 128 and 177. Fuller quotes the figures for Cambrai as 376 fighting tanks and a total, including supply tanks, of 474. Fuller, *Memoirs*, p. 187. It does seem that Amiens was a significantly bigger tank operation than any previous British effort though the German High Command must have had singularly little imagination if a tank attack on this scale was 'undreamt of'.

14. Guderian seems to have drawn on several accounts of Amiens. The opinion he forms of the significance of the battle is, however, virtually identical with that of Fuller in *Memoirs*, pp. 291317.

15. Guderian is perhaps being a little hard on the German Army here. British military documents of 1918 indicate that the British were increasingly having to take into account German anti-tank defences, with more German artillery being deployed forward in the direct fire role and an increasing use of anti-tank mines – though the Tank Corps believed it could overcome these obstacles. See for example Elles to CIGS, 19 March 1918 (PRO) W.O. 158/865.

16. Most modern authorities on chemical warfare would agree with Guderian that chemical weapon use generally tends to slow the pace of operations. Guderian here offers an important clue as to why

the Germans did not use these weapons in the early stages of the Second World War. E. M. Spiers, *Chemical Warfare* (Macmillan), 1986, p. 66.

17. The last sentence is very strange. So far from being intimidated, in the spring of 1917 the German High Command decided upon unrestricted submarine warfare, knowing that there was a high probability that this would bring the United States into the war. This did indeed happen in April and it proved fatal to German ambitions. See Fritz Fischer, *Germany's Aims in the First World War* (Norton), 1967, pp. 306-9.

The Versailles Diktat[1]

The paragraphs of Part V of the shameful Treaty of Versailles were conceived in a spirit of hate. We are no longer bound by them, but it is a salutary exercise to call them to mind every now and then. The German Army, as permitted by their provisions, was small and incapable of further development. But the most irksome feature was not the numerical weakness, or the obligation of twelve-year service. It was the prohibition of all modern weapons.

The field army was forbidden to possess heavy artillery, with the exception of a few heavy fortress guns and naval and coastal artillery, which opened the way for a measure of cautious experimentation. But the air forces, the tank forces and the U-boat forces were destroyed and forbidden outright,* and the possession of chemical weapons was likewise prohibited. The German Army was reduced to 21 regiments of infantry, eighteen of cavalry and seven of artillery, with a few weak auxiliary units. Essentially it had been reduced to a police force, incapable of conducting even a colonial war in modern conditions.

*(Note: Article 171 of the Treaty of Versailles stipulates in Paragraph 3: *Germany is likewise forbidden to manufacture or import armoured vehicles, tanks, or similar machines which may be turned to military use.*)

In accordance with the Peace Treaty of 31 August 1919 the German National Assembly passed a relevant law of implementation, where we read in Paragraph 24:

Punishments of up to six months' imprisonment, detention, or fines of up to 100,000 marks will be imposed on whoever acts contrary to the provisions of the Peace Treaty in Germany in the following respects:

1. . . .

2. . . .

3. Manufactures armoured vehicles, tanks or similar machines, which may be turned to military use.)

In terms of armament and equipment the army represented scarcely any advance on that of 1914. Most striking of all was the large number of regiments of cavalry in proportion to those of infantry and artillery. After the Armistice the enemy had all the time they needed to make the peace conditions just as disagreeable and shameful as they could, and it is unlikely that they had our best interests at heart when they came to determine the composition of our army. There could be no mistake: Germany was forced to accept an army that not only lacked any kind of offensive potential, but was incapable of putting up a sustained defensive. The only tactics that matched our military capacity would have been those of a 'fighting retreat', and, given our chronic weakness in ammunition as well as manpower, this would have degenerated in a matter of days into a disorderly flight.

It is true that the army retained its old defiant, warlike and offensive spirit, as befitted its glorious traditions. This was all well and good, and was a credit to its commanders, especially Colonel-General von Seeckt. But we were forbidden just those weapons that had shown the greatest importance and striking-power in the last war, and, lacking an everyday acquaintance with them, the army was in real danger of letting them slip from its mind altogether, or at least under-valuing them to a greater or lesser extent.

The heavy artillery could still function to a certain degree, as we have noted. The Luftwaffe and the U-boat service, although relatively new arms, had a body of officers that had come into being in the course of the war, and a tradition that was already several years old. Gas countermeasures could and did go forward.

It was much more difficult for the tank forces. In the war we had effectively denied ourselves an armoured 'service' as such, for our forty-five machines were too few to constitute anything of the sort. All that remained were the scanty experiences which a number of individuals carried in their heads, and with a few exceptions these men left the army as a result of the reductions. Up to 8 August 1918 we had shut our eyes to what tanks had achieved up to that time, and indeed the directions in which they might develop. After the war the progress of armour in foreign countries was for years on end concealed from us altogether, or at best we gained only fragmentary glimpses. Our peacetime manoeuvres were innocent of tanks or anti-tank weapons. When, finally, canvas mock-ups appeared at exercises, they had to be pushed or carried by troops against the infantry and artillery. They looked frankly comical, which was not the way to convey the image of a kind of deadly enemy, or persuade the other arms to do anything about altering their tactics, which reverted increasingly to those of 1914. We had experienced periods of reaction even after wars we had won, like the Franco-Prussian War of 1870-1, whose lessons were incorporated only in the Drill Regulations of 1888. But never had regression been more marked than after 1918.

In awareness of these dangers the Germans furnished their dummy tanks with motors. Since, however, the Peace Treaty allowed the army literally 'one' tracked vehicle, we could simulate a tank attack with any appearance of reality only on particularly favourable and obstacle-free terrain; in other words mostly on drill squares. The self-propelled mockups worked to the extent that they at last persuaded officers and men to devote a little thought to anti-tank defence, with the result that we introduced wooden guns which were supposed to represent anti-tank guns. How unpretentious we had become! I remember how proud we were when we made the tin turrets of our 'tanks' traversible, and were able to simulate machine-gun fire by a little blank firing machine. What joy we took in our first smoke generator! But our greatest secrets were our forbidden Rübezahl Tractors [called after the mythical giant of the Riesengebirge in Silesia] which were based on the clanking commercial tractor. With this machine we essayed our tank company tactics under conditions of the greatest secrecy at Grafenwöhr.

In those years only a few officers devoted themselves to a detailed and genuinely professional study of the development of armour in its tactical and technical aspects. They were limited almost exclusively to the motor transport branch [*Kraftfahrtruppe*], and what a tiny circle they were! But how we prize the memory of that time when we wore the rose-red piping of that branch of service! How we recall all the work we put in, how we strove after knowledge, our search to identify the likely development of this new weapon, the weapon of the future! In those years was laid the foundation of the discipline, comradeship and soldierly and technical proficiency on which alone the German mechanized and tank forces could have arisen when, finally, we were free of the restrictions on our armament. We who wore the rose-red piping have every right to be proud of such a groundwork: we recall with gratitude the men who in those years of trial carried forward the tank arm and its development, and prepared so effectively for its present rise.[2]

Notes

1. The Germans called Versailles a *Diktat* rather than a Treaty because they said it was imposed unilaterally rather than freely negotiated. For discussion of this point see A. J. P. Taylor, *The Origins of the Second World War* (Harmondsworth), 1964, p. 52.

2. As explained in the editors' introduction, the panzer forces grew out of the Motor Transport Troops. Guderian was originally asked to explore motorized warfare by General Tschischwitz, the head of this branch of the Army. See Guderian's memoirs, *Panzer Leader* (Arrow), 1990, pp. 19-24.

Post-War Developments Abroad

While Germany laboured under the dictates of that infamous Treaty of peace, our former enemies retained their full freedom of action. 'The weapons which brought us victory are in a state of constant activity. Tanks and aircraft show almost daily advances.' (General Débeney, in *Revue des deux Mondes*, 15 September 1934.) Here we will sketch in the technical and tactical developments affecting the various kinds of armoured vehicles, together with the weapons and resources of anti-tank defence. On this basis we can begin to discuss the future development of the devices in question, see how they shape up against one another, and evaluate their roles in the context of the army as a whole.

We will break down the treatment as follows:
– the technical development of the main types of armoured vehicles;
– the evolution of tactical concepts in those armies that have the most significant tank forces, and finally
– the present state of anti-tank defence.

1. Technical Developments

The characteristics of armoured vehicles ought to correspond to the way we intend to employ them. We will categorize and describe them accordingly:
(a) By far the greater number of armoured vehicles should be destined for combat – combat against conventional forces, but more especially against anti-tank weapons and enemy armoured vehicles. We call these machines 'tanks' (*Panzerkampfwagen*). Under this designation they are categorized either according to weight as light, medium or heavy tanks, or, since weight is a fairly arbitrary and vague definition, more usefully by their primary armament as machine-gun-, light-, medium- or heavy-gun tanks. Tanks must be capable of crossing difficult ground, and of giving their crews protection at close range at least against small-arms fire, and at medium range against anti-tank guns. They must possess all-round traverse for their primary armament, good vision, easy transportability and sufficient speed for their purpose.
(b) Armoured reconnaissance vehicles (*Panzerspähwagen*) are used for scouting, and they must accordingly have a higher turn of speed than tanks. However they must possess a reasonable degree of cross-country mobility,

and this must be all the greater if they are destined to work closely with the tank units. For operational reconnaissance, where speed is at a premium, a common solution is wheeled vehicles with two to four sets of wheels and all-wheel drive (as in the *Automitrailleuses de découverte*). Tactical reconnaissance has a frequent need for off-road and cross-country mobility, which is answered by half-tracks or wheel-track convertibles (*Räderraupen-fahrzeuge, Automitrailleuses de reconnaissance*). Combat reconnaissance is carried out in immediate association with the fighting units, and demands fully-tracked vehicles with complete cross-country mobility.

(c) Special tasks require appropriately specialized vehicles. This has given rise to amphibious tanks for swimming across water, radio tanks or command vehicles for signals and the transmission of orders, and bridge-laying and mine-plough tanks for the engineers.

Rather than burden the reader with long-winded explanations we recommend him to study the table on p. 148.

We can see even at first glance the progress in design between the tanks of 1917 and those of 1937, comparing for example the British Mark V which fought at Amiens (No. 7) with the heavy Vickers Independent (No.16), or the French Saint-Chamond (No.11) with the Char 3 C (No. 17). It calls to mind the similar advances in warships and aircraft, which have acquired greater sleekness of line, simplicity and serviceability, becoming in the process more technically 'beautiful'.

The innards have improved in step with the externals. The most diverse designs of running gear offer far greater endurance than at the time of the last war; they can be used on hard-topped roads, and they have made tanks largely independent of special transporters. Suspensions have also shown a notable improvement, which eases the strain on the crew and provides a more stable platform when the tank is firing. Engines are more powerful. The British Mark V, for example, had a 150hp engine; the modern Vickers Independent weighs about the same, at 32 tons, but has a 350hp engine, yielding improvements in climbing ability as well as speed.[1] This opens the prospect of a freer tactical deployment of armoured forces, and of bringing operational objectives within their reach – it had been the earlier limitations on range which as late as 1918 had brought many of the more high-flown schemes of our enemies down to earth. Since the war, the effectiveness of armour protection has increased many times over in respect of thickness, conformation, and the quality of the steel. All armoured vehicles worthy of the name are completely proof against small-arms fire, and most gun tanks can also withstand the smaller anti-tank guns. The contest between armour and gun has been taken at least as far among the tank forces as among the navy and the air forces.

Since the war the emphasis in the armament of tanks has been less on the number of weapons that can be taken on board, than on the performance of the gun, its suitability for the narrow dimensions of the tank, and the way it is mounted. We need only to compare the arcs of fire of the sponson gun of the British Mark V (No. 7) and the bow-mounted gun of the Saint-

Chamond (No. 11) with the turret guns of the Vickers Independent (No. 16) and the Char 3 C (No. 17) which give us the 360-degree arc of fire that is really needed. Gun sights likewise have been greatly improved by good optics.

Visibility is still not ideal, but at least it is better than before, thanks to optical provision for the driver, and improvements in the form of the slits to resist the entry of shell splinters and bullet splash.[2] All of this gives the crew better protection against injury. Most of the larger tanks now have special command cupolas which free the tank commander from having to serve the gun, and give him what he needs for effective control – especially of the larger units – namely unrestricted vision of the whole vehicle and over a field of 360 degrees, independently of the orientation of the main turret. Considerable use is made of movable mantlets to secure the necessary field of vision for the commander; periscopes have to serve instead in smaller tanks which do not have commanders' cupolas.

Communication between the members of a tank crew is effected by lights, speaking tubes, internal telephones and other devices. For external communication nearly all command tanks have radio transmitters and receivers, while all other modern tanks have radio receivers; the company commanders of the World War, hastening ahead of their tanks on foot or horseback, are figures of the past. The continuing development of radio apparatus is of great relevance to the direction of larger tank formations and their deployment for tasks in depth.

Armoured reconnaissance vehicles have made comparable progress. In the last war the chassis used to have two fixed axles, mostly with rear-axle drive only; the tyres were of solid rubber, and the total weight of the vehicle frequently approached the limit of what the chassis could bear. Reconnaissance vehicles of that type were usable only on firm roads, and they were therefore very sensitive to obstacles. This was a fundamental deficiency which meant that they were not much good at carrying out the various reconnaisance tasks which fell to their lot, and made them virtually useless on the shell-torn battlefields of the Western Front. We encountered them when the French were holding off the German offensive against the Chemin-des-Dames, and when the British sent them forward in pursuit at Amiens on 8 August. The Germans did not employ them at all.

The improvements after the war were concerned most immediately with the driving properties and cross-country mobility in particular. A number of directions were followed – two-axle drive, the introduction of a third and later a fourth axle (No. 30) with appropriate drives, the swinging half-axles, and bullet-proof pneumatic tyres. In many models steering was extended to all wheels, and heavy armoured reconnaissance cars were equipped with the facility of additional steering from the rear. Spare wheels were attached on a rotating mounting on a dead axle, which helped to prevent the hull from grounding on uneven terrain. Auxiliary tracks (No. 30) helped in climbing obstacles and crossing soft ground. The facility of driving on both wheels and tracks gave rise to the wheeled/tracked convertible armoured vehicle (*Räderraupenpanzer*), and we finally take note of the 'hermaphrodite' half-track, a speciality of the French, in which the rear wheels are replaced by tracks, while front-wheel steering is retained.

It is true to say that that the various improvements in the chassis of armoured reconnaissance vehicles came at the right time to answer the call which now arose for operational, tactical and combat reconnaissance, but the development is far from complete. As regards their suspensions the armoured reconnaissance vehicles have undergone an evolution similar to those of their close relatives, the tanks, with the difference that armour protection mostly gives place to greater speed and range, with particular attention being paid to the signals equipment.

Inevitably the technical advances in armoured reconnaissance vehicles proceed in close and mutually beneficial association with those of commercial vehicles of all kinds. As early as the World War commercial vehicles played an important role in the transportation of headquarters officers, troops and supplies. Since the war a wide and accelerating phenomenon has been the partial or total extension of powered vehicles to forces of all kinds. This process is called the 'mechanization of the army'. The first element to be affected was the high command. Isn't it nowadays inconceivable to imagine a commanding general on horseback on the battlefield, let alone a divisional commander? The officers who experienced the facility of powered transport certainly found it very welcome. The next step was the mechanization of the signals and communications units, a significant part of the heavy artillery, the engineers and almost the whole of the logistic system. Then came the setting up of motorized machine-gun and infantry units, and of army-level transport groups which were capable of carrying all kinds of forces and equipment.

The final process was to extend the new dimension of mobility to whole weapons at a time, and above all to mechanize the one arm of service which in its existing form was incapable of answering the demands of modern war, namely the cavalry. This development was at its most thoroughgoing in Britain, where the entire cavalry was mechanized, with the exception of a few mounted regiments which were left in the structure of the infantry divisions for reconnaissance tasks. As was announced in the press in December 1935, the cavalry was going to be mechanized because the mounted cavalry divisions lacked the speed, range and striking-power that were demanded in modern mechanized warfare. The French went about things rather more slowly, and out of their five cavalry divisions two have been completely mechanized, and the others by two-thirds. In contrast the Russians still maintain a large force of cavalry, despite the huge extent of the mechanization of their army.

There was an especially urgent need to mechanize the auxiliary units which were destined to act with the armoured reconnaissance and tank units. This call gave rise to motorized experimental infantry brigades and light mechanized artillery and engineers in Britain, and to *Dragons portés*, light artillery and engineers in France, and similar elements in Russia and elsewhere.

A final impulse was the restoration of German military capacity, which presented other countries with the requirement for a mechanized anti-tank defence.

2. TACTICAL DEVELOPMENTS

The notions concerning tank forces are extraordinarily numerous and diverse, and it is hardly surprising that they have given rise to a great variety of combat and transportation vehicles, and, on the organizational side, to armoured combat units and mechanized forces of all kinds. From all of this we must try to form a picture of the development of the mechanized army of the future. The process is extremely exciting, and we shall trace it with reference to the three military powers that have been of the greatest relevance to the progress of armoured forces in Europe, namely Britain, France and Russia.

After the war the British withdrew to their island fastness and slashed the numbers of their army. They scrapped or sold off most of their combat vehicles, and retained only the newest types for exercises, and as a base for experimenting with their designs for a modern army.

The following principles have determined the evolution of the British tank forces. Britain needs its army in the first place to protect its Empire. If, however, a large-scale war broke out on the continent of Europe, the best help that could be rendered to allies would be in the shape of a small but highly mobile army, possessing great striking and offensive power. This would be more useful than sending conventional divisions of infantry or cavalry, since Britain's allies had plenty of those already. What mattered was to commit the kind of force that would express Britain's industrial proficiency, namely a comprehensively motorized and mechanized army which could move at great speed and strike to great effect. In such a modern form even a small army could represent an important, even decisive, accession of strength to an ally.[3] The tank forces would play an essential role in the new army, and this was why particular attention was paid to their development. Among other things they would, in contrast to the last war, have to reckon with a strong anti-tank defence.

Since it was impossible to predict the outcome of the contest between gun and armour, and it was conceivable that the anti-tank gun might gain the upper hand, the main emphasis of British post-war development was put not so much on armour protection as other considerations – a compact and agile tank, effective means of command and control, and the ability to deliver a mass attack by surprise at the decisive spot. It was hoped that speed of movement, and exploitation of terrain and smoke-screens would lessen the danger from the anti-tank defence, and enable the assault to be pressed home successfully. There followed the inevitable conclusion that the tank assault must be separated from that of the infantry, if not immediately and totally, at least at a very early stage in their joint attack. Again it was argued that if self-preservation dictated that the tanks must part company with the infantry sooner or later, it was better to do so in a systematic way, and come to terms with whatever the tactical consequences happened to be.

What were the advantages of the tanks acting independently, and exploiting their greater range and speed? A successful attack would bring a swift victory, which would assume considerable dimensions in breadth and depth; the enemy reserves, and most importantly the motorized or even armoured units, would arrive on the scene too late. Here was the solution to the hitherto intractable problem – how to exploit success. Breakthrough and pursuit again became a real possibility, and war would assume or maintain the character of a war of movement. The tank forces would gain not only a local, tactical importance on the battlefield, but one which extended into the operational sphere of the theatre of war as a whole.

What, on the other hand, were the disadvantages of separating the tanks from the infantry? If the tank forces got too far ahead of the other troops, or veered too far out to their flanks, the tanks might be able to win a great deal of ground, but not hold it for any length of time. Again, unsupported tanks might not be able to cope with an awkward type of defence in difficult terrain. On their side the infantry might feel at a disadvantage without immediate and constant support from the tanks, and their objectives might appear impossible, or at best attainable only at an intolerable price.

In order to overcome the first of these disadvantages, the one relating to unsupported armour, the protagonists of mechanization – General Fuller, Martel, Liddell Hart and others – advocated reinforcing the all-tank units by infantry[4] and artillery mounted on permanently assigned armoured vehicles, together with mechanized engineers, and signals, support and supply elements.

This reasoning engendered the instructions published in 1927 under the title *Provisional Instruction for Tank and Armoured Car Training, Part 2*, and also the Experimental Mechanized Brigade which was set up in the same year. This formation was composed of tanks and mechanized infantry and artillery, and was organized into a reconnaissance group consisting of: one company of light tanks; two companies of armoured scouting vehicles; a main group comprising one battalion of medium tanks, one towed motorized field artillery detachment, one light battery on self-propelled carriages, one machine-gun battalion, one engineer company and one signals company. In 1928 the brigade was designated the 'Armoured Force'. This body represented the first ever experiment with a completely modern tactical formation, which was powered by the internal combustion engine alone, and which did not possess a single horse. It was intended to secure the co-operation of the conventional arms with the tank forces, and the solution was to motorize the older elements completely – indeed totally mechanize some of them – enabling them to follow the tanks at speed on the march and on the battlefield, or at least as far as enemy action permitted. The instructions just mentioned laid down the guidelines for the deployment of tanks within the novel formation, and ensured full freedom of action for future developments. However it appears that the mechanical developments at that time did not keep pace with the conceptual ones, at least to judge by some of the difficulties which arose on exercises. As a result there was something of a reaction concerning the deployment of tanks, which was the central issue.

In 1929 on the initiative of the General Staff, two experimental infantry brigades were formed. These consisted of a light tank battalion, a mortar company, and three battalions of infantry which normally marched on foot, but which could be transported on trucks if the need arose. In other words, mechanized forces and marching infantry were combined in a formation of pretty small dimensions. A number of exercises in the following years revealed shortcomings, notably that the tanks forfeited speed as a consequence of being tied so closely to the infantry.

In 1932 an all-armoured force took the field on exercise.[5] The year 1934 saw the establishment of an armoured formation which for the first time was reinforced by complements of all the other arms to provide a tank brigade consisting of: one light and three mixed battalions; one truck-borne infantry brigade of three battalions; one armoured reconnaissance detachment of three companies; four batteries of light artillery; two anti-aircraft batteries; one signals, one engineer and one medical company, and a supply train. The command of this force was entrusted to a general who had little experience in the handling of tanks, and who showed a certain lack of assurance in the way he went about his work. Snags arose in both the scenario and the direction of the manoeuvres. The operation supposed a raid in the rear of a hostile army. The force accomplished the very considerable march demanded of it and duly arrived in the enemy rear where, however, it failed to carry out any particularly bold strokes. The command was simply too cautious, and as a result not a great deal was learned about the tactical handling of the force, and still less about the actual direction of combat.[6] All the same the lessons seemed to point in a single direction, to judge by what followed in December 1935, when the entire British cavalry was united with the tank brigade in the 'Mechanized Mobile Division', except for regiments designated as divisional reconnaissance detachments. Although the names of the old cavalry regiments were retained for the sake of tradition, the decision signified the complete transformation of army-level cavalry into an armoured force. The revolution affected not only the British regular army in England, but extended to the forces overseas, and above all the troops in Egypt.[7]

The 'Mechanized Mobile Division' comprises two mechanized cavalry brigades each of one armoured reconnaissance regiment, one motorized cavalry (rifle) regiment and one light regiment of cavalry tanks, together with the existing four-battalion tank brigade and a corresponding number of artillery batteries and supporting services. It has incorporated the bulk of the tank forces of the British Expeditionary Force in a well-articulated formation capable of operational deployment. In addition the British seem to be planning to set up further tank battalions which will be at the disposal of the individual armies,[8] and whose main task is said to be co-operation with the infantry. At the time of writing two such battalions already exist. According to the most recent reports the British intend to increase their tank forces to a total of fourteen battalions.

To sum up, the post-war developments in Britain indicate the concentration of the greater part of the tank forces, including the former cavalry regiments, into an operational formation under unified command, and we can also identify the intention to create further tank units under army command for co-operation with the infantry. If we categorize the former cavalry regiments as battalions (which matches their combat power), the Mechanized Mobile Division is made up of two reconnaissance battalions, three light and two mixed tank battalions, and two rifle battlions, with artillery and supporting services. The emphasis in the armament of the division is plainly with the tank forces.

Within the tank units we can distinguish between light armoured reconnaissance battalions, which are equipped with the necessary reconnaissance vehicles, and light and mixed combat tank battalions. In the light battalions the companies each comprise seventeen light tanks and two or three close-support tanks; in the mixed battalions they contain six medium and seven light tanks, and two or three close-support tanks.

The combination of light, medium and close-support tanks within the same company permits a very fluid conduct of the battle, with the armoured self-propelled artillery keeping up with the attack, and guaranteeing fire support for the light and medium tanks when they are engaged in close-range combat. The tank attack therefore becomes independent of the support from conventional artillery, which can follow the advance of tanks only as far as the gunners can keep their progress under observation. The way their armoured forces are arranged leads to the conclusion that the British intend to entrust their tanks with tasks extending deep into the enemy rear, and therefore to endow even the smallest units with considerable freedom of action.

The French have chosen a different path from the British in every respect. In 1918 they were relieved of the immediate threat from their eastern neighbours, and yet they retained their massive level of armaments, which they saw as a potent means of enforcing their policies on their defenceless former enemies. One of the consequences was that their tactics and their operational objectives were largely shaped by the vast quantity of equipment which remained from 1918, and its technical capacities. Thus the French tank forces retained the light Renault (No. 10) as their main vehicle – a slow, short-ranged machine which was destined mainly for immediate co-operation with the infantry. However, the potential enemy was weak and virtually devoid of anti-tank capability, and the French tactical notions seemed to promise victory within a reasonably short time.

The one immediate shortcoming was the fact that the Renault had only limited climbing, obstacle-crossing and wading ability, and was therefore unsuitable for attacking positions in difficult terrain. For decisive victory in this context the need was not so much for speedy, wide-ranging tank forces in the operational dimension, as large, heavy tanks which could meet those specific technical requirements. This was probably why at the end of the war the French took over a number of heavy British Mark Vs, and continued the development of new types of heavy tanks which had been proposed by Estienne during the war. The weight of these machines grew from 50 to 69, and again from 74 to 92 tons. The Char D could climb at 45 degrees, surmount an obstacle three metres high, cross a gap six metres wide, and wade to a depth of 3.5 metres – these are capabilities which must be taken into account by works of fortification with any claim to be secure against tanks. It hardly needs to be said that the French designated these armoured monsters as 'defensive weapons'. When the proposal was made at the Geneva disarmament conference to ban all offensive weapons, the French proposed that heavy assault tanks should be defined as beginning only at more than 92 tons.

As long as the French were dealing with a defenceless Germany, they could be confident enough about the shape their own offensive tactics ought to take: in the last war most infantry attacks had been broken by the defensive fire of machine-guns, but now the machine-guns would be tackled by the light tanks, with the mass assault by infantry following immediately behind. The strong defensive positions would be overcome mainly by the heavy breakthrough tanks, which would make the breaches for the penetration.

It soon became clear, however, that the chief threat to the breakthrough was presented by motorized reserves; that being so, motorized assault troops were needed to exploit the initial success. The shortage of military manpower in the post-war years ruled out the setting up of completely new formations, and indicated instead the partial or total transformation of the forces that were least able to meet the demands of modern warfare in effectiveness and striking power, namely the cavalry. In about 1923 the French began experiments to convert their cavalry to a motorized and up-to-date force. This process developed in a number of directions, which are not always easy for us to trace.

Operational or long-range reconnaissance was clearly beyond the capability of mounted scouts. The answer lay in the multi-wheeled armoured reconnaissance vehicles which were produced by the firm of Berliet. Armoured reconnaissance vehicles were also suitable for tactical, or close-range reconnaissance when the rearward wheels were replaced by tracks, which increased cross-country mobility. Such half-tracks or hermaphrodites are represented by the Citroen-Kégresse and the Panhard-Kégresse (No. 28), which are of particular significance in the development of French armoured vehicles. They are also used to transport the motorized infantry, the *Dragons portés*, who are intended to support the armoured reconnaissance vehicles.

After a period of evolution which extended over several years, the French experiments finally produced in 1932 the new Type 32 Cavalry Division, which consists of two mounted and two motorized brigades, in addition to the armoured reconnaissance units. As far as we can ascertain this type of division consists of:

Divisional headquarters, with an aerial and photographic detachment;

Two cavalry brigades of two regiments apiece, each regiment comprising one headquarters squadron, four mounted squadrons, one machine-gun squadron, and one supporting weapons squadron;

One motorized brigade of one tank regiment, and one regiment of three battalions of *Dragons portés*;

One artillery regiment of two light detachments and one heavy detachment;

Engineers, signals units, anti-tank weapons and supporting services.

The division's tank regiment, listed above, is composed of a motorized reconnaissance detachment of motor-cycles and twelve armoured reconnaissance vehicles, and a tank detachment of twenty reconnaissance vehicles and twenty-four reconnaissance tanks, producing a total ratio of combat to reconnaissance vehicles of 24 : 32. The division numbers some 13,000 men, 4,000 horses, 1,550 motorized vehicles and 800 motor-cycles.[9]

The constitution of the Cavalry Division was tested in the course of several years of exercises and large-scale manoeuvres. Despite all the assurances that had been given by the enthusiasts of the noble horse, the combination of animal and engine proved of more harm than benefit for military purposes. If the motorized element went ahead of the other forces they won ground which was of theoretically great use for the attack and established early contact with the enemy; but then they had to wait a long time, often too long, for the mounted brigades to come up, and before that happened the precious ground was liable to be given up again, frequently with large losses of valuable equipment. A number of officers concluded that the advantages of powered vehicles must be exploited to the full, and they urged repeatedly that motorizing the entire division would make it far more effective. A more immediate way of bringing together horse and engine was to send the cavalry brigades out first, and hold back the motorized brigade in reserve until the focal point of the battle had been identified and the brigade could intervene. In practice the procedures turned out to be too complicated for the short distance which the motorized force often had to cover, and the cavalry units could have done the same more effectively. In addition the pace of the whole division was tied to that of the horse.

These were the reasons why as early as 1933 the French had begun experimenting with a fully mechanized division, and the result was the *Division légère méchanique*. We do not have precise details of how it is organized, though in outline it is made up of the following elements:

Divisional headquarters, with auxiliary elements and aerial detachments;

One armoured regiment for reconnaissance;

One tank brigade for combat;

One dragoon brigade on motor vehicles;

One artillery regiment of two light detachments and one heavy detachment;
Engineers, signals and rear service personnel.

The total strength stands at about 13,000 men and 3,500 powered vehicles (including 1,000 motor-cyles). The division has about 250 tanks, of which approximately 90 are for combat and the rest for tactical and operational reconnaissance.[10] In contrast the Type 32 Division has 56 tanks, of which 24 are for combat.

The Type 32 Division has itself undergone detailed reviews, which have resulted in a second Type 32 Division being converted to a fully motorized division, with a third division to follow in 1937.

We have seen enough to recognize that the divisional establishments of reconnaissance vehicles are very high in proportion to those of combat vehicles. It is fair to conclude that the *Division légère méchanique* is destined primarily for reconnaissance tasks and is unsuitable for serious combat – an imbalance which is the legacy of the origins of the divisions in the cavalry. It is an open question how long this state of affairs will last, but the weakness of the structure has undoubtedly been recognized, as is shown by a speech by the Minister of War Daladier, announcing that in 1937 France intended to experiment with a heavy tank division – in other words one that was capable of offensive action. Daladier made his standpoint clear in the following passages:

'In addition to the army of the people – the conscript army – do we not stand in need of a professional army or a special force of tank divisions, composed exclusively of long-serving soldiers? There are some who envisage this body as a force capable of immediate action, a shock army. Others welcome it as a means of shortening the term of active conscript service, or eventually doing away with it altogether.

'But at root we all strive for speed and striking power.

'It has already been mentioned from this platform how in 1933, in full agreement with the High Command, I created the first *Division légère méchanique*. A second is in the process of formation, and a third will follow.

All three divisions will consist of fully trained men and have the necessary means of transport fully under their command.

'It is my belief that the light divisions must be supplemented by a number of heavy divisions. We will undertake a number of very important experiments with this division at the end of the coming summer [1937].

'We need a much more specialized army. We must have various kinds of divisions for various tasks. In all of these momentous questions I am in full agreement with the High Command, which is just as determined as I am to furnish the French Army with everything that modern technology has to offer.' (*France Militaire*, No. 16, 565-6.)

Nearly all armies cling to the notion that operational reconnaissance must be carried out by cavalry divisions or by their successors, the light mechanized divisions. But is not this concept outmoded or plainly wrong? The original cavalry divisions were never intended exclusively for reconnaissance. When Napoleon I, their creator, set up divisions of cuirassiers, dragoons and light cavalry, he destined the first two exclusively for combat roles, and only the light divisions primarily for operational reconnaissance. In the nineteenth century European cavalry was intended, organized and trained for decision in battle – an objective which was rarely attained, since breech-loading rifles appeared in the meantime and made the cavalry incapable of winning battles by cold steel. In the campaigns of 1866 and 1870-1 cavalry achieved very little in the shape of operational reconnaissance, perhaps because of its imbalanced peacetime training. Only after the failure of cold steel in the battle-winning role did there arise the desire and (because of the attachment to sword and lance) also the need to find new purposes for cavalry.

Operational reconnaissance was recognized as one such task – and there was something to be said for it before aircraft and tanks were invented. But it is questionable whether there was any need to employ whole divisions or even whole cavalry corps for this purpose, especially when they were organied in such a way that every element was capable of the reconnaissance role, but not a single one of them had enough combat potential, by which we mean firepower, to break any kind of serious resistance. It might have been better to destine only part of the cavalry regiments for reconnaissance, with the appropriate training and equipment, leaving the greater part of the division for the combat role. This notion would have led to the constitution of specialized reconnaissance regiments with light armament and a small complement of vehicles, but good means of communication, and distinct combat regiments and brigades with plenty of heavy armament, plenty of ammunition and a sufficiency of artillery. In all likelihood the cavalry in the World War would have been better at combat and reconnaissance than it actually was. Perhaps the cavalrymen should have rid themselves of the idea that reconnaissance was their private kingdom, and paid more attention to building up their combat power – which again might have resulted in the formation of more solid and usable cavalry divisions before the war.

If we follow our chain of reasoning to the present time, we must raise the question whether it has been a good idea to fit our large mechanized formations primarily with reconnaissance vehicles, at the expense of their effectiveness in serious combat. Our reservations appear all the more justified since operational reconnaissance must fall chiefly to the air forces, since they can reach more deeply into the enemy rear and work more speedily than ground reconnaissance. Operational ground reconnaissance should therefore be seen as a supplement to aerial reconnaissance. Europe is a relatively small theatre of war, and this work can therefore be carried out by small, but fast-moving and combat-capable reconnaissance forces, which can receive rapid support in case of need from the mechanized combat formations.

It is clear from Monsieur Daladier's address that the development of equipment and the fitting out of their forces have progressed sufficiently far

Specifications of Tanks, from Heigl, *Taschenbuch der Tanks*

Photo No.	Designation of Tank	Country	Crew	Armament		Ammunition	Armour/mm	Speed in km per hr
				Gun	M.G.			
1	Heavy Mark I, 1916	GB	8	2 x 57mm	4	-	5-11	5,2
2	Heavy Mark V, 1918	GB	8	2 x 57mm	4	2,000 gun & 7,800 m.g.	6-15	7,5
3	Heavy Schneider, 1917	FR	6	75mm	2	96 gun & 4,000 m.g.	5,4-24	6
5	Light Renault FT, 1917	FR	2	37mm	or 1	240 or 4,800 m.g.	6-22	8
6	Heavy St. Chamond, 1917	FR	9	75mm	4	106 gun & 7,488 m.g.	5-17	8,5
7	Medium Mark A Whippet, 1918	GB	3	-	3	5,400 m.g.	6-14	12,5
8	Medium Vickers Mark II, 1929	GB	5	47mm	6	95 gun & 5,000 m.g.	8-15	26
9	Medium A 7 V, 1918	GER	18	57mm	6	300 gun & 18,000 m.g.	15-30	12
10	Light L K II, 1918	GER	4	-	1	3,000 m.g.	bis 14	18
11	Heavy Vickers Independent, 1926	GB	10	47mm	4	-	20-25	32
12	Heavy Char 3 C, 1928	FR	13	1x155mm 1x75mm	6	-	30-50	13
13	Light Renault N C 2, 1932	FR	2	-	2	-	20-30	19
14	Medium T 2, 1931	U.S.	4	47mm	1x12mm 1x7.6mm	75 gun, 2000 & 18,000 m.g.	6,35 bis 22	40
15	Light Mark II, 1932	GB	2	-	1	4,000 m.g.	8-13	56
16	Light Renault U E	FR	2	-	1	-	4-7	30
17	Light Carden-Lloyd (Amphibious)	GB	2	-	1	2,500 m.g.	bis 9	9.7 in water otherwise 64
18	Fast Christie	RUS	3	47mm	1	-	6,35 bis 16	110 on wheels 62 on tracks
19	Light Fiat Ansaldo, 1933	ITY	2	-	1	4,800 m.g.	5-13	42
-	Airmobile Carden Lloyd (Russkii)	RUS	2	-	1	-	6-9	40
22	Armed Recce. Vickers Guy	GB	6	-	2	6,000 m.g.	6-11	50
23	Armoured Recce. Panhard-Kégresse-Hinstin '29	FR	3	37mm	1	100 gun, & 3,000 m.g.	5-11,5	55

Performance											
Km range on one fuel tank	° of climb	Climbs m	Pushes over tree of cm	Crosses gap of m	Wades depth of m	Weight	Horsepower	Length	Width	Height	Ground clearance
24	22	1,20	to 50	4	1,00	31	105	8,6	3,9	2,61	0,45
64	to 35	1,20	to 55	4,50	1,00	37	150	9,88	3,95	2,65	0,43
75	30	0,40	0,40	1,80	0,80	13,5	60	6	2	2,40	0,40
60	45	0,60	to 25	1,80	0,70	6,7	40	4,04	1,74	2,14	0,50
60	35	0,40	0,40	2,50	0,80	23	90	7,91	2,67	2,36	0,41
100	40	0,80	0,35	2,50	0,90	14	90	6,08	2,61	2,75	0,56
220	45	08,0	0,40	2,00	1,20	13,4	90	5,31	2,74	3,00	0,45
80	25	0,40	-	3,00	0,80	30	-	7,30	3,05	3,04	0,50
-	45	0,90	0,30	2,00	1,00	9,5	60	5,70	2,05	2,52	0,27
320	40	1,50	0,76	4,57	1,22	30	350	9,30	3,20	2,75	0,60
150	45	1,70	0,80	5,30	2,00	74	1980	12	2,92	4,04	0,45
120	46	0,60	0,25	2,10	0,60	9,5	75	4,41	1,83	2,13	0,45
145	35	-	-	1,80	1,20	13,6	323	4,88	2,44	2,77	0,44
210	45	0,58	0,30	1,52	0,75	3,6	75	3,96	1,83	1,68	0,26
180	38	0,40	-	1,22	0,70	2,86	35	2,70	1,70	1,17	0,26
260	30	0,50	-	1,53	amphibious	3,1	56	3,96	2,08	1,83	0,26
400	40	0,075	0,20	2,10	1,00	10,2	343	5,76	2,15	2,31	0,38
110	45	0,60	-	1,50	0,90	3,3	40	3,03	1,40	1,20	0,25
160	45	0,40	-	1,22	0,66	1,7	220	2,46	1,70	1,22	0,29
220	-	-	-	-	-	9,25	75	6,58	2,35	2,86	0,25
200	35	0,40	-	1,20	1,20	6	66	4,75	1,78	2,46	0,25

to enable the French to proceed with full-scale field trials with a heavy tank division – in other words a division furnished mainly with battle tanks.

A number of authorities have emphasized for a considerable time that the technical development of the tank must be expressed in tactical and operational terms that match its capability, and the French too will come to recognize this truth. The Future will surely triumph over the Perpetual Yesterday, and more specificially over post-war training regulations which were compiled under conditions which no longer obtain. The *Division de choc* proposed by Charles de Gaulle in 1934 (*Vers l'armée de métier*, Paris) is approaching reality.[11]

We are sufficiently acquainted with the industrial performance and military aptitude of the French to be in no danger of underestimating our western neighbours. It would therefore be as well to reckon before long on the appearance of heavy French tank divisions, built around a core of modern tanks, equipped with light, medium and even heavy artillery, and possibly the necessary motorized support in the shape of reconnaissance, infantry, gunners, engineers and auxiliary services.

Our review of the French armoured forces leads to the following conclusions. The armoured equipment remaining from the last war was technically primitive and would have been usable only in close co-operation with the infantry and on good ground. It predicated also a likely enemy who was denied anti-tank weapons, tanks and motorized battleworthy reserves. As long as these conditions prevailed the French were sure of winning through, even if they attacked in a slow and methodical way, and their pace was tied to that of the infantry. The only snags might have been caused by encountering natural or artificial obstacles that were beyond the capacity of the light Renault tanks. In such cases the French intended to use an appropriate number of heavy breakthrough tanks.

The rearmament of Germany changed the picture almost overnight. The predominance of the French tanks came to an abrupt end, and the French had to reckon first with serious anti-tank defence, then with hostile tank forces, and finally with large self-contained motorized and mechanized formations. This was a death blow to the theory and practice of keeping the tanks bound to the infantry, and scattering them more or less evenly among the attacking units. Where was the sense in spreading tanks over the frontage of the offensive when anti-tank defence could spring up anywhere, taking a toll of the armour when the attack was confined by the terrain into a few narrow avenues of approach? Was there not a need for speed if one was to exploit success and prevent the enemy from focusing their forces on the threatened spot and counter-attacking?[12]

The French Minister for War and the high command were therefore acting quite sensibly when they converted their cavalry to mechanized forces. Proceeding with the same logic, they are now concentrating their heavily armed and heavily armoured tanks into heavy armoured divisions, the *Divisions de choc*.

'Once tanks became considerably faster than infantry, the limited concept of tanks as infantry support was gradually replaced by the notion of large mechanized formations. They are not confined to breakthrough battle tanks in the narrow sense. They also incorporate reconnaissance elements, and cross-country transport to carry the necessary minimum of infantry and artillery immediately behind the tanks, so as to secure the ground which the armour has conquered. The newly acquired characteristic of the tank, its speed, can be exploited at the first blow. It is now possible to think of employing large mechanized formations independently.

'Here is something new to match the tactics of the present day, and here also lies the possibility of restoring mobility to warfare.

'The large mechanized formations are the real instrument of the offensive . . . Striking-power and speed open a new world of possibilities.' (Lieutenant-Colonel Lançon, in *La France militaire*, 1937, No. 178.)

If it is a question of breaking an enemy force in full defensive posture, then in future the heavy tank divisions will open the way for their lighter brothers, and the motorized and horse-drawn formations will follow. If, however, there is a considerable distance between the two belligerents, and it is a matter of enveloping and turning movements, the light tank divisions can hasten ahead and secure important features of the ground, and fix and hold up hostile troop movements and strike at the enemy communications, thereby facilitating the advance and deployment of the heavy tank and motorized divisions. In any case the opening encounters will be carried out in co-operation with tanks, where the localities are suitable, and as the battle goes on the armoured forces will increase in significance rather than diminish.

At the beginning of 1937 French possessed rather more than 3,000 light and heavy pieces of artillery (excluding fortress and anti-tank guns) and more than 4,500 tanks; which means that the number of tanks exceeds by a wide margin the number of guns, even in the peacetime army. No other country shows such a disproportion between armour and artillery. Figures like these give us food for thought!

The tank forces in Russia developed on different lines from those in Britain and France. In the World War the mighty Russian army had no tanks; it did not have the indigenous industry to manufacture its own machines, and geographical isolation prevented it importing tanks from its allies. Only in the Civil War did a few tanks fall into Russian hands. The absence of armour meant that in the war against Poland a large force of cavalry was able to play a decisive role under the energetic leadership of Budenny, admittedly against an enemy who was badly led and had little defensive capacity.

After the Civil War Russia addressed itself with urgency to the task of building an indigenous armaments industry. This process inevitably extended over many years because of a total lack of direction and expertise; but now it must be regarded as largely complete. Over the same period the Russians studied the progress that foreigners were making in every field of technology, and how they might copy it. Tanks and their auxiliary weapons also came under this scrutiny.

The Russian practice was to buy and test the leading models of foreign tanks, and then build reproductions adapted as necessary to Russian conditions and requirements.[13] The same freedom from tradition and technical preconceptions is shown in the way the Russians have developed their tactics. As far as we can judge from outside, every one of their twenty-three armoured corps has a regiment of battle tanks, quite apart from further regiments which stand at the disposal of higher levels of command. A number of auxiliary forces have also been set up – motorized infantry divisions and rifle brigades, towed and self-propelled artillery, reconnaissance and other motorized units – though it is not yet possible to identify permanent groupings in large formations.

On the other hand we can form a reasonably good picture of how the Russians intend to employ these modern forces of theirs. The evidence comes from military literature and the reports of their various exercises. 'A decisive success', in the words of Kryshanovsky, 'is attainable only through the simultaneous destruction of the main enemy deployment to its entire depth, in both tactical and operational dimensions. This demands the action of strong, fast-moving forces which possess great striking power and mobility.' (M. J. Kurtzinski, *Taktik schneller Verbande*, Potsdam.) The Russians seek to put the principle of the simultaneous destruction of the whole deployment into practical effect through the way they will arrange their 'motor mechanized' forces for the attack. To this end they distinguish between three kinds of formation:

1. NPP = Immediate Infantry Support;
2. DPP = Long-range Infantry Support;
3. DD = Long-range Forces.

The NPP Forces are built around the 6-ton Vickers-Armstrong-Ruskii AT 26 tank, which is armed with a 59mm gun and two machine-guns, and is proof against armour-piercing small-arms ammunition. Twenty-six of these tanks at a time will provide cover for 35 machine-gun tanks, the Vickers-Carden-Loyd-Russky T 27, which is lightly armoured but has good climbing capability. The establishment of NPP tanks is completed by twenty light BA 27 tanks with a 37mm gun, and a few little 'Bronieford' tanks. The designation 'Immediate Infantry Support' sums up the purpose of these forces. To fulfil their mission, however, they need the protection of tanks of a more formidable kind, which can smash through strong positions and eliminate the artillery and anti-tank defences. This work is assigned to the DPP formations.

The Long-range Infantry Support Forces are built around heavy breakthrough tanks (Types M I and M II), which have a 75mm gun as their primary armament, together with one or two armour-piercing guns of smaller calibre and some machine-guns. For their light armour the DPP forces are equipped with a number of Vickers-Armstrong-Russky 6-tonners and Vickers-Carden-Loyd amphibious tanks.

Once the DPP and NPP forces have succeded in breaking through the enemy front and pinning down the defenders, the DD forces will exploit and, with plentiful air support, advance against the enemy command centres, reserves, lines of communication and rearward installations. For this purpose they have a particularly fast tank, which they have adapted from an American design, namely the Christie-Russky 34 (No. 23) with its 47mm gun and single machine-gun. The armour is pretty thin, but the vehicle has a range of 400 kilometres, and a speed of 110 kilometres per hour on wheels and 60 on tracks. Altogether the Christie-Russky is a particularly well-designed and tested machine. In addition the DPP units are equipped with a large number of armoured reconnaissance vehicles and 6-wheel Ford amphibious variants armed with 37mm guns and machine-guns.

Basically there is something to be said for the way the Russians have organized their forces: rapid and long-ranged tanks to operate in the enemy depth; heavily armoured tanks with heavy guns to do battle with the enemy tanks, anti-tank weapons and artillery on the main battlefield; light tanks mostly armed with machine-guns to clear the infantry combat zone. On the other hand the triple classification of tasks demands a whole inventory of specialized tanks, with all the attendant disadvantages.

The number of Russian tanks is put at 10,000 machines, and the armoured reconnaissance vehicles at 1,200. These are figures of an impressive order, not least because the armour will be acting in co-operation with a powerful and modern air force,[14] and it will gain added potency if the Russians manage to bring their road and rail network into a usable condition. In 1936 large-scale exercises were held in the Belorussian and Moscow Military Districts. The object was to test the co-ordination of the motor-mechanized forces with the infantry and cavalry divisions, but especially with the air force, which for the first time carried out major descents of air-landed troops in the enemy rear under the protection of paratroops – seeking to tie down enemy reserves, or to complete envelopments by the ground forces. On the same occasion a number of specially adapted aircraft transported and landed light armour.

A number of foreign powers have taken up the idea of paratroops and air-landed forces from the Russians. Opinions are as divided as to their usefulness as they are on the subject of tanks. Some authorities treat them as a joke; others say that Central Europe is so densely populated that air drops and landings will be pinpointed, brought under attack and neutralized in short order. But, as with all innovations in the field of military technology, it is unwise to jump to conclusions before undertaking a serious examination of the pros and cons of new forces and the necessary countermeasures. Otherwise there will be some painful surprises in store when it comes to real combat.

Russia possesses the strongest army in the world, numerically and in terms of the modernity of its weapons and equipment. The Russians have the world's largest air force as well, and they are striving to bring their navy up to the same level. The transport system is still inadequate, but they are working hard in that direction also. Russia has ample raw materials, and a mighty armaments industry has been set up in the depths of that vast empire. The time has passed when the Russians had no instinct for technology; we will have to reckon on the Russians being able to master and build their own machines, and with the fact that such a transformation in the Russians' fundamental mentality confronts us with the Eastern Question in a form more serious than ever before in history.[15]

We have seen how since 1918 the tactical development of armoured forces in the three most important military states of Europe has followed the technical development only step by step, and has sometimes lagged behind. It proved difficult – especially in official circles in Britain and France – for the authorities to free themselves from conceptions inherited from the past, or hammered home during four years of static warfare. Not infrequently the forces of reaction proved to be stronger than the urge to progress. No wonder that the smaller states, with their limited resources, chose to wait and see how the organization and deployment of armoured forces shaped up elsewhere. These states are therefore irrelevant to our purpose. It is a different story with the development and present state of anti-tank defence, which have an immediate bearing on our investigations.

3. ANTI-TANK DEFENCE

Germany's decision not to build tanks relieved the enemy Alliance of the need to provide for anti-tank defence in the years from 1916 to 1918; the Germans on their side underestimated the importance of the new tank weapon in general, which meant that they too neglected anti-tank defence. The consequence was that Germany was beaten.[16]

Having identified the causes of that defeat, Germany's first concern was to cast about for means of defence against its potential enemies – they were armed with tanks and they were lurking on every side. We therefore looked into the whole question of anti-tank defence, and took a number of practical measures as a result. In its turn the restoration of German military sovereignty – and with it the certainty that Germany would acquire tanks – has meant that for several years now the other powers too have had to look hard at defence against tanks. We will first examine the basic principles.

Nature herself offers the most effective protection against tanks, though she does not necessarily offer it in every location and at all seasons of the year. Steep slopes, broad and deep waters, ditches, and dense and tall woodlands may present absolute barriers to tanks; the ground behind such obstacles is termed 'tank-proof' (*Panzersicher*). Less pronounced obstacles will make the movement of tanks more slow and difficult; built-up areas too may have the effect of holding up tanks, and providing good protection behind walls and in houses and cellars. Such terrain is described as 'restricted for tanks' (*Panzerhemmend*). Open, lightly wooded country with variable cover tends to facilitate the armoured attack, and is called 'favourable for tanks' (*Panzergünstig*).

The defenders will strive to exploit 'tank-proof' terrain for their own purposes, siting their defences there, or using it to support one or both of their flanks. 'Restricted' terrain is easier to find; it is a considerable asset to anti-tank defence, and it increases the time in which the anti-tank weapons can take their effect. Sometimes the engineers will be able to make conditional obstacles into absolute ones, for example by scarping slopes and ditches, creating inundations and swamps, making abatis of sufficient height, breadth and depth around the edges of woods. Artificial obstacles can be established in open terrain in the form of iron rails embedded in concrete, and stakes, concrete walls, concrete pyramids, rolls of wire and mines.

However Nature is not always lavish in providing 'tank-proof' or even 'restricted' terrain; the availability of time, labour, materials and munitions will limit the extent of artificial obstacles and how far we can adapt the ground, and the effectiveness of such barriers is in any case conditioned by how well they are concealed and what kind of resistance they can present to obstacle-clearing parties. Also the military task and situation might leave the defenders with no option but to make a stand in terrain that favours tanks. All of this means that defending troops need armour-piercing weapons, otherwise they might find themselves in a situation like that of the German infantry in 1918, defenceless and faced with an impossible task. Even if frontal attacks are ruled out, the enemy might still break through on a neighbouring sector and present a sudden threat to a flank. Effective anti-tank weapons are consequently an essential part of the equipment of all forces, but the infantry in particular.

Anti-tank weapons can be accounted completely effective only if they act speedily enough to smash the enemy attack before it can reach the main line of infantry defence. If the attack is contained at a later stage it will only be at the cost of heavy casualties among the infantry, and perhaps also their complete destruction – a case of 'the operation was successful but the patient died'. Infantry anti-tank weapons must therefore be light and compact enough to be used in the front line, but sufficiently responsive and long-ranged to halt the attack in the way just outlined.

We will take some figures as an example. Enemy tanks are launching an attack at a speed of 12 kilometres per hour, covering 1,000 metres in five minutes. Let us suppose that the defenders have a gun which can get off eight aimed rounds a minute at an effective range of 600 metres; placed at the forward edge of the infantry combat zone, such a weapon can fire 24 rounds before the tanks break into the position; this figure rises to 40 rounds for a weapon with an effective range of 1,000 metres. The corresponding figures for an automatic weapon, firing in short bursts, are 100 rounds per minute instead of 8, which produces 300 rounds at an effective range of 600 metres, and 500 at an effective range of 1,000 metres. Calibre for calibre, the automatic weapon's greater volume of fire must nevertheless be bought at the price of heavier and bulkier gun and ammunition.

In the British peacetime army the fourth battalions of their infantry brigades have motorized anti-tank companies equipped with sixteen automatic cannon of 20mm calibre. France has introduced automatic 25mm cannon, and every infantry battalion is to be assigned three guns of this kind towed by tractors (No. 21). Germany has 37mm guns (No. 43) towed by a 6-wheeled powered vehicle. A variety of lighter anti-tank weapons of 12mm and greater, the so-called 'anti-tank rifles' (*Tankbüchsen*), are currently under development; served by a small detachment, they are of approximately the same dimensions as machine-guns but are capable of piercing armour only at short range. It seems that no final decision has yet been taken whether to introduce weapons of this sort into service, though the British are conducting trials with a 12mm anti-tank rifle which weighs 16 kilograms, has a rate of fire of between six and eight rounds a minute, and is effective against light armour up to 450 metres. (*Militär-Wochenblatt*, 1937, No. 48.)

It would be wrong to conclude from this that infantry are now able to overcome the threat from armour. It is well within the capacity of military technology to produce tanks that give full protection against weapons of such a calibre, and yet have full motorized mobility and do not exceed the load limits of road bridges. Such tanks already exist, notably in France (see table on p. 148). The weapons mentioned above will be ineffective against an assault by heavy tanks of this kind, and if the enemy choose to put heavy tanks in their first wave they will not only wipe out the light anti-tank defences, but allow a swarm of light tanks to come forward; these, being invulnerable to armour-piercing rounds of small-arms calibre, will finish off the infantry as well as complete the breakthrough.

This danger may be countered by introducing anti-tank guns of heavy calibre. They are already under development. Indeed we have a report (*Militär-Wochenblatt*, 1937, No. 46) of a British vehicle-drawn, Vickers 75mm anti-tank gun of this kind. The carriage permits an all-round traverse, and the gun fires a 6.5kg round with an initial muzzle velocity of 595 metres per second and a muzzle pressure of 117 metric tons. The duel between artillery and armour has existed for some time in naval and siege warfare, and it has now spread to tank combat, and it will extend inevitably to war in the air. But, just as we should not stop building warships, fortresses and aircraft simply because of the armour-piercing capacity of heavy and super-heavy guns, so the undeniable strength of anti-tank defence does not mean that that the building of tanks has become of no purpose on this account. If we fell in with that line of argument we would have to agree that General Douhet's conclusion was valid generally, and not just for Italy, when he argued that air forces alone retain the capacity for offensive action, and land forces will have to content themselves with standing on the defensive. Douhet's views have in fact been hotly contested, and especially by those who stress that land warfare is as decisive as ever it was, and that land forces still have adequate offensive capability. If we wrote off armour as a battle-winning instrument we would simply be back in the position of the Germans between 1916 and 1918, when they renounced the use of tanks. In fact it would spell the end of successful offensive action in ground warfare.

Let us return to the contest between tank and gun, and apply it to our own circumstances. We must do everything we can to promote anti-tank defence, and work just as hard to guarantee successful counter-attacks through the instrument of powerful tank forces of our own. Logically, from the standpoint of the defender, we need guns and ammunition capable of combating the most formidable types of tank known. Our existing heavy ordnance has range and penetrating power in plenty, but it lacks the mobility and speed to meet a surprise attack by heavy breakthrough tanks; it is also deficient in rapidity of response and traverse, and its sighting mechanism is unsuitable for tracking fast-moving targets. The current medium and heavy artillery therefore has a reasonable, but by no means assured, chance of success against heavy breakthrough tanks. The lesson is that we need new types of guns.

Mines are another valuable asset of anti-tank defence. They can be laid at short notice in the requisite breadth and depth, and they can be readily concealed in any fairly broken ground. This kind of defence will be by no means easy for the enemy tank forces to identify, and they will suffer heavy losses if they do not carry out proper reconnaissances, and have not been able to clear lanes through the minefields by artillery fire or mine-clearance parties. All of this makes the mine a formidable enemy to the tank.

The use of mines is restricted by the stocks available, and still more by the precautionary measures which the existence of minefields inevitably imposes on the defending troops. Extensive minefields are in fact a severe limitation on the defenders' freedom of movement. Irregular groupings of mines and the scattering of individual mines avert this danger in the short term, but they still represent a danger to our own troops. The actual location of the mines is usually known only to the engineers who have laid them. The other arms are generally less well informed, especially if there are frequent changes in the troops who are holding the positions in question. This disadvantage becomes all the more marked for the defenders in the case of mobile warfare, even when they are on the retreat and do not intend to consolidate on a new line.

To sum up, anti-tank defence is founded on the natural or artificial obstacles which the ground presents to armour, on the additional obstructions that we lay across the area of the attack in the form of mines, and on the fire of anti-tank guns of various calibres. We can therefore distinguish between two main categories of anti-tank defence:
– rigid, immobile forms which are conditioned by the ground, or are strongly emplaced;
– mobile weapons that are designed to meet the situation in almost any terrain.

Both forms must be exploited. In the sectors where we intend to limit ourselves to the defensive, the static types can be employed in systematic barriers of the kind we find in modern defensive systems; the mobile forces can act as reinforcements and mobile reserves for the static defences, but in addition they will be able to move swiftly to offer a defence wherever it might be needed in unfortified ground. The static forms emphasize terrain, systematic reconnaissance, and the availability of labour, materials and time; the mobile forms depend on the existence of suitable defensive forces, but will spring into action whenever and wherever the call comes.

Anti-tank units and engineers are the most suitable forces for anti-tank defence, but for combat against unarmoured troops they must be backed up by the firepower of machine-guns – and if necessary artillery as well – and by reconnaissance and communications elements. Thus, in addition to the anti-tank weapons which are incorporated permanently in the divisional structure, there may also be blocking units (*Sperrverbände*) standing at the disposal of the high command. Bearing in mind the nature of the attack, the prospect of a successful defence may well hang upon the speed with which these forces can move up and get into the fight. It is almost impossible to exaggerate the value of troops of this kind – fast, endowed with strong defensive combat value, and specially trained and equipped for their work.

Notes

1.Guderian was clearly impressed with the Vickers Independent. It had a 3pdr gun in an all-round traverse turret and four machine-guns in sub-turrets. It was intended to be capable of operating without close infantry support and it could achieve 20mph. It looked impressive but had a number of design faults. It was never put into mass production for financial reasons. David Fletcher, *Mechanized Force* (HMSO), 1991, pp. 24-5.

2. 'Bullet splash'. This is the phenomenom of the interior of a tank being spattered with bullet fragments. These could inflict extremely painful wounds. David Fletcher, *Landships* (HMSO), 1984, p. 16.

3. Guderian's argument here about the best kind of Field Force for Great Britain to adopt is similar to that put forward at one stage by Basil Liddell Hart – a small but highly mechanized force would be best. The five-division Field Force which the British General Staff planned from 1935 onwards was to be quite highly mechanized by the standards of other armies at the time and was to include one armoured division. But the Army got a very low priority in British rearmament and from January 1938 to February 1939 the General Staff was operating under what amounted to a ban on preparations specifically intended for a campaign on the Continent. The development of British armoured forces suffered accordingly. Liddell Hart was one of those in Britain most opposed to the preparation of a Field Force for the Continent in these years. The standard work on the British Army between the wars is Brian Bond, *British Military Policy Between The Two World Wars* (Oxford), 1980.

4. In this context Guderian is, perhaps, somewhat over-generous to Fuller and Liddell Hart – crediting them with his own logic. Neither of them was as convinced of the importance of having infantry closely integrated with tanks as was Guderian himself. The Experimental Mechanical Force (EMF) of 19278 was actually the idea of George Lindsay – a Royal Tank Corps (RTC) officer whom Guderian does not mention – though Fuller played a significant role in establishing it and Liddell Hart commented on it at length in the *Daily Telegraph*. Lindsay and Fuller were insistent on keeping conventionally armed infantry out of the force though they wanted to include a battalion of motorized machine-gunners. One particularly interesting feature of the EMF was the inclusion of 'Birch guns' – self-propelled 18pdrs. The self-propelled gun concept in the British Army went back to the First World War when the Tank Corps ran Gun Carrier tanks. Post-war interest in the concept by the Royal Artillery was, however, short-lived. See J. P. Harris and F. H. Toase, *Armoured Warfare* (Batsford), 1990, pp. 327 and Winton, *To Change An Army* (Brassey's), 1988, p. 92.

The EMF experienced difficulties in the co-operation of different arms that had differing mobility characteristics. After 1928 the thinking of Fuller and some other leading British tank enthusiasts became increasingly tank-centred. They played down the need for co-operation with other arms. Fuller arrived at a particularly extreme and unsound position in his major theoretical work on armour, *Lectures on F.S.R. III* (Sifton Praed), 1932. Guderian does not appear to have been familiar with this, though he used Fuller's memoirs which were published later. For a discussion of Fuller's thought at this stage see B. H. Reid, *J. F. C. Fuller: Military Thinker* (Macmillan), 1987, pp. 160-3. Reid largely dismisses the common belief that Liddell Hart was more perceptive than Fuller in seeing the need for intimate co-operation between infantry and tanks.

5. Guderian is wrong about the year. The 1st Brigade RTC, an experimental formation, was assembled, under the command of Charles Broad, during the training season of 1931. See Harris and Toase, op. cit., pp. 37-40.

6. Guderian's account of the Mobile Force exercise of 1934 is generally accurate but he gets one fact badly wrong. The Mobile Force was commanded by George Lindsay, one of the most prominent and experienced RTC officers of the day. Guderian is probably getting confused with the EMF of 1927 which was commanded by an infantry brigadier – Collins, who had no previous armour experience. The disappointing showing of the 1934 Mobile Force is attributable partly to the ground rules for the exercise having been laid by Major-General 'Jock' Burnett-Stuart in such a way as to make its life as difficult as possible, partly to unfair umpiring, and partly to a dispute between Lindsay and his most important subordinate commander – Hobart. The best account is Winton, *To Change An Army* (Brassey's), 1988, pp. 174-83.

7. The decision to create a mechanized Mobile Division to replace the horsed cavalry division was taken by Field Marshal Sir Archibald Montgomery-Massingberd, Chief of the Imperial General Staff, late in 1934. The decision to mechanize the entire cavalry of the line was set out in a major document (PRO – W.O. 32/4612) prepared under Montgomery-Massingberd's direction, in September 1935. Entitled 'The Future Reorganization of the British Army', this document attempted to map out the army's development for the next five years. See Harris and Toase, op. cit., pp. 42-4.

8. Guderian was right. In addition to their Mobile (armoured) Division the British set up 'Army Tank Battalions' equipped with slow but heavily armoured 'Infantry Tanks' for close co-operation with infantry advancing on their feet. Liddell Hart, *The Tanks*, Vol. I (Cassell), 1959, pp. 337-8.

9. The French persisted with hybrid tank-cavalry divisions. They still had five of these – of a type called *Divisions Légères Cavaleries* – in 1940. Guderian was correct in predicting that their combat power would be negligible. Robert Icks, *Famous Tank Battles* (Profile), 1972, pp. 102-14.

10. Guderian is here discussing what the French called *Divisions Légères Mécaniques* (DLMs), the nearest thing they had to panzer divisions. In 1940 there were still only three, with a fourth in the process of creation. Richard M. Ogorkiewicz, *Armour* (Stevens), 1960, pp. 64-7.

11. Guderian was partly right. The French did belatedly create *Divisions Cuirassées de Réserve* (DCRs). These were armoured divisions which included some very heavy tanks – Char Bs. But the divisional structure was faulty, the number of tanks inadequate and the Char B (despite its impressive size, armour and armament) had design faults one of which, a one-man turret, was also found in the Somua, the best tank held in the DLMs. There were three DCRs in May 1940 with a fourth, commanded by Charles de Gaulle, in the process of formation. The first three were sent against the German bridgeheads over the Meuse but proved very ill-adapted to long road marches, arrived in bits and were destroyed piecemeal by the seven panzer divisions in that sector. Gunsburg, *Divided and Conquered* (Greenwood Press), 1979, pp. 68-9.

12. Guderian's logic is compelling. The French, however, still had more than half of their tanks split up in small units for co-operation with infantry divisions in 1940. See R. Icks, op. cit., pp. 102-3.

13. A particular example of the greater technical sophistication Guderian is here discussing is the Russian acquisition of a prototype designed by American inventor, Christie, which incorporated his revolutionary suspension system. The US Army could not make this system work. British tanks based on it (like the Crusader) tended to be unreliable. The Russians made it work extremely well and used it in many tanks including the excellent T34. Ogorkiewicz, op. cit., pp. 191-2 and 225-9.

14. Guderian is here somewhat over-estimating the Red Air Force. Despite its enormous size it proved neither very powerful nor modern and was totally outclassed by the Germans in 1941. It was not until 1943 that the Soviets were beginning to achieve air parity on their front. Alexander Boyd, *The Soviet Air Force* (Macdonald and Jane's), 1977, pp. 88-124 and 167-84.

15 Guderian is extremely prophetic here. His insight may derive partly from a visit to Russia in 1933 during which he saw a Soviet tank factory. His claim in his memoirs that he opposed Operation 'Barbarossa' must be taken seriously in view of the evidence of this passage. Guderian, *Panzer Leader* (Arrow), 1990, pp. 142-4.

16. This is, of course, a gross over-statement of the importance of tanks in the First World War, made for effect.

The German Mechanized Forces

1. The Era of the Dummy Tanks. Military Sovereignty

Germany's mechanized troops did not spring into being fully formed like Pallas Athena from the forehead of Zeus. On the contrary, their evolution was a long-drawn-out story of deprivation under the restrictions imposed by the Treaty of Versailles, and they had to contend – and contend still – with the opposition provoked in our own camp through their sheer novelty and unfamiliarity.

Under the old War Ministry (*Reichswehrministerium*), the Inspectorate of Mechanized Forces (*Inspektion der Kraftfahrtruppen*) was the only organ of the army responsible for the concept of mechanization as a whole, and consequently of maintaining the tradition of the weak German tank forces of the World War. In addition to the work of mechanizing the army in general, the efforts of the Inspectorate extended in two directions. The first was to explore the business of transporting troops by truck. A number of exercises were arranged to this end, beginning with the Harz Exercise of 1921. This was directed by the current Inspector, Major-General von Tschischwitz, and concerned the movement of a single battalion. In later years a number of reinforced battalions and regiments were put through a large number of trials – lengthy cross-country marches, and shorter ones in the course of manoeuvres – and in the process the Inspectorate learned a great deal about how to prepare and manage large-scale transport by truck.

The second direction involved the setting up of a core of tank troops, albeit under conditions of the greatest difficulty. The Allies permitted us to have nothing more than the so-called 'armoured trucks for the transportation of personnel' – by which the French and British understood a goods truck with a sheet iron superstructure. After long negotiations they finally permitted the armoured vehicle depicted in Plate 31, which had vertical sides and no traversible turret or permanently mounted weapons. By devising a chassis with 4-wheel drive and rear steering we were able to build a vehicle which had a certain value for dealing with internal disorders in Germany, and, as we shall see, was usable in training. The armour proved to be too heavy for the chassis, without being proof against the rifle-calibre armour-piercing round, and we were so short of resources that we were unable to build even the limited number of vehicles allowed by the terms of that disgraceful treaty. All the same we put it to use in the first of our officer training courses, when the machine saw service in a number of small-scale exercises, especially in reconnaissance work. We acquired some useful lessons as a result. Here, indeed, was the first spark of inspiration for developing the motorized forces into a proper tank arm, and it was never going to be extinguished.

Since the Treaty of Versailles forbade us to manufacture tracked vehicles, we had the idea of building multi-wheeled vehicles which would have a certain amount of cross-country mobility. We accordingly began to devise a number of 8- and 10-wheeled machines. However it was specified at the same time that the vehicles must be amphibious, which led to some very complicated and bulky machines that were subject to a great number of teething troubles. In spite of efforts extending over several years it proved impossible to build vehicles of this kind that could be of any use in war.

The same period saw the first projects for tanks with caterpillar tracks. Development went ahead slowly, however, because the emphasis in the training of the mechanized troops had to shift to logistic transportation.

A later Inspector, General von Vollard-Bockelberg, recognized how inefficient it was to keep on training all the diverse elements of the mechanized forces as a single body – tanks along with motor-cycles, columns of trucks, ambulances and so on. Instead he instituted a separation of functions, though the work inevitably had to proceed under conditions of the greatest secrecy, on account of the restrictions imposed by the Treaty of Versailles.

Mock-ups were mounted on Hanomag chassis, and several companies of 'tanks' were formed, and duly went through their paces with this forbidden weapon under the astonished gaze of the other troops on exercises. Clumsy though they were, they showed the need for effective anti-tank weapons, and kept alive a debate of sorts concerning tank operations and anti-tank defence.

Motor-cyclists were concentrated in a Motor Cycle Rifle Company (*Kraftradschützenkompagnie*), and in the manouevres of 1928 they were used for the first time in association with the existing 'tanks' and mechanized infantry.

Courses were held for the officers of the mechanized forces, and they extended beyond technicalities to the actual tactics, and the co-operation of the mechanized forces with the other services. It was not long before limited numbers of officers of those arms were also admitted to the courses.

In this and the following years the officers of the *Kraftfahrlehrstab* (Motor Transport Instructional Staff) not only established uniform tactical and technical principles within their own branch of the service, but engaged in detailed discussions which laid down the guidelines for a rebuilding of the German tank forces – or at least as far as this was possible through theoretical studies and our practical exercises with the mock-ups. In later years the *Kraftfahrlehrstab* became the foundation of the *Kraftfahrlehrschule* (Mechanized Forces Instructional School) of the new army.

General von Stülpnagel continued the work of organizing and training along the lines that had been pioneered by his predecessors. The motorized detachments were reinforced by a number of motorized squadrons comprising one company each of motor-cyles, armoured reconnaissance vehicles, 'tanks' and anti-tank troops. Inevitably they were all still equipped with dummy vehicles and wooden guns (Plate 32).

On 1 April 1931 the former chief of staff, General Lutz, was appointed Inspector of Motorized Forces. His earlier career had been in the technical branches, and during the war he had functioned as commander of the motor transport of one of our armies.

On instructions from army command, General Lutz conducted two sets of three exercises each on the respective training areas of Grafenwöhr and Juterbog. The purpose was to train a dummy tank detachment in working with a reinforced infantry regiment, and to gain some experience of anti-tank defence. These six exercises provided a useful stimulation for the later build-up of our tank forces in general, as well as helping to set the specifications for our future tanks and given the immediate impetus for a number of design projects. Foreign literature and machines underwent careful examination, as a means of tapping the experience of the other countries that had been constructing tanks for the last sixteen years. All the same our later prototypes were not immune from developmental problems, because one cannot replace long experience of building real tanks by imitation or work on the drawing-board.

In the autumn of 1932 four motorized reconnaissance detachments and a composite motor-cyle battalion took part for the first time in the large-scale exercises. The organization stood up very well, and the efforts of our new forces and their officers earned considerable applause. This cheered us no end, and we pushed ahead with the work of development.

However there remained one obstacle – the fact that our military and political leadership lacked the nerve to free itself from the Versailles restrictions. As far as our own arm was concerned this decisive step was accomplished almost overnight, when executive power was transferred to Adolf Hitler on 30 January 1933. The armour plates of our 'mock-ups' became visibly stronger, and were now proof against the attentions of the street urchins who used to amuse themselves by boring holes through them. The wooden guns were banished. The reconnaissance detachments were increased to four companies each, anti-tank detachments were organized in three companies, and we began experiments with motorized infantry and tanks.

By 1 July 1934 the experimental work had attained such dimensions that it became necessary to set up a special Command of the Tank Forces (*Kommando der Panzertruppen*), which was entrusted to its first commanding general, Lieutenant-General Lutz, the former Inspector. The task of the new command was to continue the experiments with the mechanized forces, and explore and test the tactical structures that might put these new formations to the most effective use. In the autumn of 1935 the various cogitations and practical exercises culminated in large experimental manoeuvres at Munsterlager, of which the most important result was the decision to establish three panzer (armoured) divisions. These were formed on 16 October 1935 under the overall authority of the Armoured Forces, and command of the individual authorities was entrusted respectively to Lieutenant-General Baron von Weichs, Lieutenant-General Fessmann and myself. The tank and anti-tank forces, the motorized infantry and the reconnaissance detachments as a whole were constituted as a new arm of the service under the designation of Motorized Combat Forces (*Kraftfahrkampftruppen*).[1] We shall now turn to the individual branches.

2. Armoured and Motorized Reconnaissance

The purpose of reconnaissance is to provide the commander with an accurate assessment of what the enemy is doing; in effect information of this kind furnishes the basis for command decisions. Reconnaissance is categorized according to the means of collecting intelligence into aerial reconnaissance, ground reconnaissance, signals intelligence (telephonic, wireless and so on), information obtained through spies and other means. The various agencies of intelligence are complementary, and, if one fails, another must take its place. Military reconnaissance is further subdivided into the operational, tactical and combat levels. Operational reconnaissance is a long-range affair which is at the service of the high command, and is carried out primarily through the air forces. However the air forces are unable to determine unconditionally whether such and such an area is occupied or not. Good enemy camouflage, night and fog, bad weather, extensive mountains and forests and large built-up areas can all render reconnaissance difficult or downright impossible. Aerial reconnaissance is incapable of keeping up a constant surveillance, or of maintaining contact with the enemy. Aerial reconnaissance has undeniable advantages – its difficulty of interception, its speed and its great range – but it cannot dispense with the need for good ground-based reconnaissance.

It it is to be of any use, intelligence must reach the commander with the minimum of delay, and speed and security are at a premium. That is why the horse has been replaced by the motor vehicle, and especially for operational and tactical reconnaissance. By definition the scouts must keep ahead of the forces who are supposed to follow them. Mounted reconnaissance is therefore suitable only for infantry divisions, and even here there is a growing demand for motorized reconnaissance, because of improvements in the cross-country mobility of vehicles.

Motorized ground reconnaissance is carried out by armoured reconnaissance vehicles. Operational reconnaissance demands considerable range and speed on the part of the vehicles, considerable combat power in terms of armament and armour, and long-range wireless apparatus. Since operational reconnaissance is conducted mainly on roads, the preference is for wheeled vehicles, which are given a measure of cross-country mobility by multi-wheel drive and rear steering. In closer contact with the enemy, reconnaissance is accomplished by light armoured vehicles or motor-cycles. Half-tracks or wheel-track convertibles are suitable for tactical reconnaissance, which calls for greater cross-country mobility, and combat reconnaissance is answered mainly by tracked vehicles. Most of the armoured reconnaissance vehicles have armoured-piercing guns (Plates 33, 34, 35).

Several armoured reconnaissance vehicles together make up an armoured reconnaissance troop (*Panzerspähtrupp*), though the actual strength and composition is determined by the task, and engineers, motorized infantry and heavy weapons may be attached as the need arises. The armoured reconnaissance troops seek out and maintain contact with the enemy – by night as well as day. Until the main force engages they report by dispatch rider, telephone or wireless. It is sometimes said that the crews of armoured reconnaisance vehicles are effectively blind and deaf. That simply is not true. When they are close to the enemy the reconnaissance forces advance by bounds from one observation post to the next, keeping their eyes and ears open all the time; if necessary they will drive up to good viewpoints, and they will leave their vehicles if this helps them to catch sounds better, especially at night. Good drivers will always keep their vehicles well concealed, and they are careful not to barge straight into anti-tank positions. The engines of modern tanks make no more noise than the clatter of hooves, and they are certainly quieter than neighing horses. Armoured reconnaissance vehicles and forces are incomparably better than cavalry in respect of combat power, speed and facilities of communication, and they will run short of fuel only when their commanders do not know their job. At present their most immediate weakness is that they do not have full cross-country mobility; but that is something which will come in time.

A certain number of light and heavy armoured reconnaissance troops form an armoured reconnaissance company (*Panzerspähkompanie*). Several armoured reconnaissance companies, with infantry on motor-cycles or in vehicles, together with heavy weapons and engineers, make up a reconnaissance detachment (*Aufklärungsabteilung*). The headquarters of the armoured reconaissance detachment acts as a sustaining and directing agency. It helps the reconnaissance detachments to keep going for several days on end, by providing reliefs and drawing on its reserves, and it has the ability to project them along a new avenue of advance which may come as a surprise to the enemy.

The reconnaissance detachments are responsible for seeking out and reporting as much as they can, without drawing attention to themselves. They must be speedy and agile, have a good range, possess good means of communication, and be responsive to command. The smaller they are, the better they will be at their work. Their combat power, and in particular their armament and protection, must be adequate to enable them to prevail against their enemy counterparts. If their task demands something heavier in the way of fighting capacity, this must be forwarded to them as necessary.

In combat the armoured reconnaissance troops and detachments act mostly on the offensive, which is the best way to destroy the reconnaissance capability of the enemy and enhance our own. We should exploit any opportunities which arise to inflict damage on the enemy, as long as fighting can be combined with the primary mission of reconnaissance. If heavier vehicles are not available, modern armoured reconnaissance vehicles have sufficient firepower to be employed in a variety of combat roles – pursuit, covering a retreat, throwing out screens in front of the other forces, and securing the flanks and rear.

Altogether our reconnaissance detachments are an excellent instrument for conducting operational reconnaissance over considerable distances for the high command or the army headquarters, and they are just as good at providing tactical reconnaissance for armoured divisions, other motorized formations or indeed any forces that are transported on vehicles. As the first elements to come in contact with the enemy, the reconnaissance detachments must be organized in peacetime just as they will be in war. There will be little opportunity to play around with them if we are overtaken by sudden developments, and, if commanders, troops and the auxiliary arms are thrown into action without having settled down together, this would put a question mark against our ability to carry out reconnaissance at the outset of hostilities, when it is so important. In other words it would be nothing short of a crime. This is a consideration that ought to outweigh all the supposed interference with training; the problems here are trivial – we have coped with them so far, and we will do so in the future. In any case they weigh heavily only with commanders who are unfamiliar with the work of motorized reconnaissance.

The reconnaissance detachments were the first of the four branches of our motorized forces to come into being after we regained our military sovereignty, and they are particularly close to our hearts. They have established the conduct of reconnaissance on modern principles which answer the peculiar needs of the armoured forces very well; indeed they are an essential component of those forces – qualifying in virtue of their ancestry, equipment, armament, training and leadership.

3. Anti-Tank Detachments

After we had created serviceable armoured reconnaissance detachments, our next priority was to form an effective defence against enemy armoured reconnaissance vehicles and tanks. This task involved all arms of the service.

The infantry regiments were given an integral anti-tank defence in the form of a fourteenth company which was armed with 37mm guns; the cavalry too were equipped with this weapon. The engineers in their turn developed anti-tank mines, and a variety of obstacles which they created from combinations of wire, stakes, abatis, inundations and ditches. The artillery set to work improving its capability against tanks at all ranges by learning to position its guns more effectively and by new fire tactics. A whole system of anti-tank defence came into being, and the need now arose to extend it in depth and create anti-tank reserves which would be put at the disposal of the higher levels of command. The Inspectorate of Mechanized Forces had already conducted highly successful experiments with a 37mm gun on pneumatic tyres (Plate 43), and it now had the task of setting up highly mobile and, consequently, motorized anti-tank detachments for the commanders in question. These detachments were duly furnished to all the larger formations of the army. They developed their tactics in co-operation with the armoured forces, and they were able to enhance the defensive capacity of the army as a whole against one of its most dangerous enemies, the tank.

The anti-tank detachments (***Panzerabwehrabteilungen***) have the responsibility for providing protection for their parent formations whether they are at rest, on the move or in action. This also helps to preserve the defensive capability of the other arms, by relieving them of the need to divert assets of their own to anti-tank defence. By themselves or, better still, in association with the engineers, the machine-guns and artillery, the anti-tank detachments also provide a means of halting unexpected enemy armoured thrusts, containing breakthroughs and checking envelopments and outflanking movements, and so giving our command the time to devise appropriate countermeasures. Forces assigned to this work are called 'blocking units' (***Sperrverbände***).

Leading the anti-tank detachments is a demanding business. On the one hand we have to site them in good time in locations from where they can use the full range of their guns to protect the forces or places for which they are responsible; however they must not betray themselves before the enemy tanks put in an appearance, and they need a degree of cover against hostile artillery. In addition their fire positions ought to be on tank-proof or at least tank-restricted ground. If they are unable to keep their firepower intact to surprise the tanks when they actually arrive, or if they are caught by the tanks when they are on the move, the whole defence will be compromised.

The commanders can do a great deal to facilitate the work of anti-tank defence by making intelligent choices of the rest areas, approach routes and above all the gun positions. One can dispense with anti-tank guns in tank-proof or tank-restricted ground, and concentrate them instead to defend terrain that offers good going for armour. As far as their equipment, time and labour allows, the engineers must also play their part by accentuating natural obstacles, and securing dead ground by barriers. However the engineers can't do much in this respect while in the offensive role when protection against enemy counter-attacks must rest entirely with the anti-tank guns; during the advance the guns must be kept sufficiently far forward to be able to consolidate whatever gains have been made.

The penetrating power of the anti-tank rounds is of crucial importance for the success of the defence. If the attackers' tanks are impervious to most of the fire aimed at them, they will overcome not only the anti-tank guns but ultimately also the defending infantry and engineers – and this latter task can be accomplished even by the light tanks of the enemy rear echelons. It is a different story when we have guns that can pierce all the available tanks of the attacking forces, and when those guns are sited at the decisive place at the right time. In that case a successful tank attack will be purchased at an excessive price, or become simply out of the question.

Effective use of anti-tank guns also depends on:
(a) terrain: steep slopes and undulating ground are unfavourable for the defence;
(b) on how the ground cover changes with the season: anti-tank guns are designed to be low-lying for the sake of concealment, and good fire positions might be difficult to find in summer [because of the vegetation];
(c) on the time of day and the weather: darkness and poor light at the beginning and end of the day make aiming difficult and prevent the guns being used to their full range. Mist and rain have the same effect by obscuring the optics of sights with droplets; it is also difficult to aim into the sun;
(d) on the effects of enemy artillery fire, even when it happens to be no more than throwing up dust and smoke, and laying down smoke-screens.

The anti-tank gunners will be put to a hard test if a number of these circumstances happen to coincide, and if the enemy put in a surprise attack with massed armour. Only superbly trained and highly disciplined troops with good nerves will be able to withstand such an ordeal. We have every confidence that we possess troops of this kind.

4. Tank Forces

Our reconnaissance and anti-tank detachments were new creations which had no counterparts in foreign armies. For tank forces, however, there were a whole series of precedents in all the most important military states. We have already described how they developed during the war and afterwards in Britain, France and Russia. The Inspectorate of Armoured Forces therefore had the difficult responsibility of deciding which of the disparate foreign models he should recommend to our own high command as the most suitable for German conditions, or whether indeed to form an entirely new doctrine.

Two things at least were clear: we could not pursue British, French and Russian tactics at the same time. Secondly we could not fashion a doctrine of our own when we were entirely devoid of practical experience, and indeed had no more than a sketchy knowledge of the wartime experiences of the French and British. After mature consideration it was decided that, until we had accumulated sufficient experience on our own account, we should base ourselves principally on the British notions, as expressed in ***Provisional Instructions on Tank and Armoured Car Training, Part II***, 1927. This document was clearly set out, and it not only offered the pointers we needed for our own experiments, but opened the avenues for development that seemed to be closed off in the better-known French regulations of the period, which sought to tie down the tanks to the infantry. The high command approved, and so it was that until 1933 the intellectual training of the officer corps of the motorized troops of the future tank force was carried out according to the British regulations. By now German views on armoured warfare were being expressed more and more vocally, and they emerged partly from our cogitations and partly from what we had learned from our experiences with the dummy tank units. The German notions had much in common with thinking abroad, but also a number of distinctive features.

Leaving aside special German circumstances for the moment, it is worth asking what, in general terms, conditions the formation of doctrine? Relevant factors are the geographical situation of a country, the strength or weakness of its borders, its raw materials, its industry and the state of its armaments compared with those of its neighbours. These influences, and especially the latter, make an immediate impression and trigger off demands for a response to meet altered circumstances. When, however, an arm of the service is under development, the fundamental task is not to keep up with every passing current of opinion; on the contrary, it is important to preserve a certain detachment from the moods and trends of the moment. We have to identify clear objectives after a due process of consideration, and then maintain purposeful technical developments over the whole span of time they will need for their fruition. Continuity is attainable only when the development is invested in the same hands over the long term, and when those hands can act with the necessary authority. Unified direction is all the more important when new arms of the service are in the infancy of their technical and tactical formation, and of their equipment and training. Even in later years, when the development does not have to proceed at such a breakneck speed as at present, there will still be cogent reasons why the potential of mechanized forces will be exploited to the full only when they are constituted as a full branch of the army.

To resolve such issues it is vital to establish the basic purpose of the tank forces. Are they intended to storm fortresses and permanent defensive positions, or to carry out operational envelopments and turning movements in the open field; to act at the tactical level, making breakthroughs on our own account and checking enemy breakthroughs and envelopments; or will they be no more than armoured machine-gun carriers working in close co-operation with the infantry? Will we try to resolve an imposed deadlock by one mighty, concentrated commitment of our main offensive weapons? Or will we renounce their inherent potential for speedy, far-reaching movement, for the sake of tying ourselves to the snail's pace of the infantry and the artillery batteries, thereby renouncing all prospect of a speedy decision to the battle and the war?

The days are long past when tanks were an auxiliary of the infantry; in fact we may almost assume the contrary, since the French hold that any infantry attack without tanks is no longer practicable. But we won't go into those arguments any further.

It is patently absurd to make the conscious decision not to exploit the potential of a weapon to the full. For that reason the specifications for the ultimate development of the weapon must be set as high as seems practicable at the time. If, for example, we can possess the wherewithal to attack at speed, it seems ridiculous to force the tanks to offer slow-moving targets to enemy fire, just because old-fashioned infantry would otherwise be unable to keep up with them. Now that technology can put the infantry in armoured escort vehicles which can move every bit as fast as the tanks, it is the tank which must determine the speed of the infantry; the French have grasped this point, and they have placed their *Dragons portés* in armoured carriers. Again, it makes no sense to bring a tank attack to a halt for several hours simply to enable horse-drawn artillery to change position, when it is now technically possible to tow guns by powered vehicles or mount them on armoured self-propelled carriages, and to provide their gun detachments and forward observers with the mobility of armoured vehicles. The tanks must not follow on the artillery, but the other way around.

Tanks will lose the capacity to concentrate on the decisive spot if they are incorporated as organic elements of all the infantry divisions.[2] Many of the machines will end up in terrain that stops them or slows them down, exposing them to heavy losses, and they will be forced to accord with the slow-moving tactics of horse-drawn artillery and marching infantry. The possibility of speed is killed stone dead, and we forfeit all real hope of attaining surprise and decisive success in combat. We will find it difficult to employ tanks *en masse* – and the increasing anti-tank capacity of all armies means that the concentration of armour is still more important for victory than it was in 1917. We will be unable to hold rearward lines and reserves of tanks, and lose thereby the means of exploiting at speed any successes on the part of the first echelon. We will grant the enemy time to bring up reserves, re-establish themselves in rearward defences, beat off our enveloping movements and concentrate for counter-attacks.

We now proceed to the selection of the types of tanks, their armament, their protection, how they are organized and how they will be furnished with complementary and auxiliary weapons. These will in turn be determined by the ends for which the tanks are intended.

Tanks do not need to be particularly fast when they are destined only for co-operating with the infantry, and when we do not form any kind of armoured spearhead to deal with the enemy's defences and artillery. However they do require very strong armour because they will be moving slowly and be exposed to artillery and anti-tank fire over a considerable length of time, especially when the range closes. Their armament consists of machine-guns and at least a small-calibre primary armament, so as to provide them with a modicum of defence against enemy tanks, or artillery pieces with gun shields which might appear at long range. Infantry escort tanks of this sort are organized in small units as far as detachment level, and they are neither trained nor equipped for combat in large formations. Their senior officers become no more than advisers to the infantry headquarters, and tactical responsibility for their employment shifts to the middle and lower-ranking officers of the infantry. The result will be that the tanks will be employed in penny packets, as was the British and French armour in 1918, though with much smaller prospects of success than then.[2]

What, on the other hand, are the specifications for tanks that are intended to break into enemy positions in battle, or execute deep breakthroughs aimed at reaching the enemy comand centres and reserves and destroying the hostile artillery? They need at least a partial covering of armour that is impervious to most of the anti-tank weapons. They require greater speed and range than do the infantry-escort tanks, and an armament of machine-guns and a gun up to 75mm calibre. Their gap-crossing, wading and crushing capabilities should be sufficient to enable them to deal with field fortifications. Lightly armoured machine-gun carriers may be attached to the tank formations for the purpose of clearing the infantry combat zone; they are sufficient for this work, since most of the defending artillery will have been knocked out by our heavy tanks.

Such tank forces must be concentrated in large formations, and provided with the complementary and auxiliary weapons they need for independence of action, just like the infantry divisions. Their immediate leadership has been already trained for this work in peacetime, and the responsibility for committing them lies with the high command. They are deployed *en masse* in both breadth and depth. They strive to exploit tactical success into the operational dimension. Enemy armoured attacks will unfailingly come in the future, but we will be able to meet them in tank-to-tank combat by large formations trained for this kind of fighting. Concentration of the available armoured forces will always be more effective than dispersing them, irrespective of whether we are talking about a defensive or an offensive posture, a breakthrough or an envelopment, a pursuit or a counter-attack.

The final category of tanks covers those that are destined to storm fortresses or permanent positions. In addition to their strong armour and heavy armament (up to 150mm), they must have considerable obstacle-crossing and wading capacity and great crushing power. When one is building tanks of this kind one very rapidly arrives at weights of from 70 to 100 tons, something which only the French have so far been able to achieve. There will never be many heavy tanks of this kind, and they will be used either independently or within the structure of the tank forces, according to the mission. They represent an extremely dangerous threat and are not to be underestimated.

Germany has attached great importance to the principle of the unified leadership and training of the armoured forces. Drawing on wartime lessons, we have renounced any idea of limiting tanks to the role of infantry escorts, and from the outset we have been determined to create an arm that is trained to fight in large formations and will be equal to any task that might fall to it in the course of time. The panzer (armoured) divisions were founded in accordance with this philosophy, and they embrace tanks, and everything those tanks need in the form of supporting and complementary arms – all of them in ample number, and all of them, it goes without saying, fully motorized.[3]

Within the tank regiments, every detachment is furnished with the machine-guns and the varieties of artillery it needs to wage fire-fights at short, medium and long range, and to answer the urgent call to meet an enemy tank attack with a sufficient number of armour-piercing weapons. One last element is required to ensure that the various calibres are used to their fullest effect – for the tank brigade and regimental commanders to deploy them in a way that answers the needs of the situation, and to take care in assigning their missions.

5. MOTORIZED INFANTRY

The experience of combat in 1917 and 1918 shows that infantry and tanks can work together effectively only when they are put through frequent and thorough training in the business of co-operation. This works best and most consistently when a certain number of infantry units are incorporated permanently with the tank units in a larger formation. We have seen how, when the war was still in progress, the French placed an infantry company on permanent assignment with every detachment of tanks in preparation for the Battle of the Aisne in 1917. The 17th Light Infantry Battalion played this role in the first attck on the Chemin-des-Dames; likewise two battalions of dismounted cuirassiers were attached to the tanks for the assault on the Laffaux angle (see p. 69). Of course at that time the infantry followed their tanks on foot, since there were no cross-country vehicles available, and the objectives of the attacks were limited. Nowadays the French have introduced the dragoon brigade on armoured half-tracks as a component part of the *Division légère méchanique*. The aim is the same. It is clear that ever since tanks appeared the French have adhered to the notion of permanent infantry support for tanks, and have followed this line quite logically in developing special formations for operational deployment. Two things are needed to accomplish a tank attack at speed and exploit and expand any success without wasting time – equipment and means of transport that answer that call for speed, but also specialized tactical training and constant practise.

We do not possess armoured cross-country vehicles for transportation.[4] Those of our infantry units that are destined for co-operation with tanks are therefore moved partly by motor-cycles and partly on cross-country trucks. The motor-cyle infantry have already performed well on reconnaissance in association with armoured reconnaissance vehicles, and they can be used for a whole variety of functions, since they are speedy, easy to conceal, and can make their way along any kind of road and across any terrain that is not altogether too difficult. We have plenty of good motor-cycles in Germany and replacements present no problem. The truck-borne infantry are protected against the elements, and in addition to the men and their equipment the vehicles carry extra loads such as ammunition, entrenching tools and engineering requisites, together with rations for several days. The present trucks are too bulky to be ideal; they have difficulty in negotiating narrow roads with sharp bends, and they are hard to conceal.

As we have already indicated, the main tasks of motorized supporting infantry are to follow up at speed behind the tank attacks, and exploit and complete their successes without delay. They need to put down a heavy volume of fire, and require a correspondingly large complement of machine-guns and ammunition. It is debatable whether the striking power of infantry really resides in the bayonet, and more questionable still in the case of motorized troops, since the shock power of tank formations is invested in the tanks and their fire power. The French have drawn the appropriate conclusion and have equipped all their infantry companies with sixteen light machine-guns each, as opposed to the nine of their German counterparts. Combat is not a question of storming ahead with the bayonet, but of engaging the enemy with our firepower and concentrating it on the decisive point.

According to General Field Marshal Graf von Moltke firepower has an offensive character: 'On occasion it can act as an agent of total destruction, deciding the issue on its own account.' (*Verordnungen für die höheren Truppenführer*, 24 June 1869.) He could maintain even in those days that front-line infantry firing at speed could defeat even the most desperate enemies. In his own words: 'The bayonets of the charging troops are powerless against them; the attackers' rifles might be every bit as good, but what use can you make of them when you are in movement and incapable of controlling your weapon?' (Ibid.) Von Moltke's perception is now eighty years old, but it has still not completely sunk into the consciousness of our army. In 1913, just before war broke out, the German infantry looked on machine-guns as no more than an auxiliary weapon: 'Here we must put on record the warning that the value of this weapon should not be over-estimated; we must not follow the error of the French in 1870-1, when they acclaimed the *Mitrailleuse* as all-conquering. The most decisive arm in battle is the infantry. When riflemen are in dire trouble, let alone in something less than dire trouble, they must not fall into the bad habit of looking for help from their supporting weapons – the machine-guns. Rather they must find in themselves the resources to surmount their difficulties.' (*Vierteljahrshefte für Truppenführung und Heereskunde*, 1913, 314). Nowadays we hear the same 'warnings' directed against the call to augment the number of machine-guns and, it hardly needs to be said, against our own views on tanks.

What we desire is a modern and fast-moving force of infantry, possessing strong fire power, and specially equipped, organized and trained in permanent co-operation with tanks.

Notes

1. For a concise modern account of the rise of the panzer forces in Germany see Harris and Toase, *Armoured Warfare* (Batsford), 1990, pp. 51-69.
2. This was precisely the problem with many of the tanks the French Army deployed in 1940. See Icks, *Famous Tank Battles*, (Profile), 1972 pp. 102-3.
3. This last is one of the crucial paragraphs in the book and gives the clearest indication of the reasons for the superiority of German over British armour in the Western Desert. The unified leadership and training of the panzer divisions embracing both tanks and all the complementary arms had no parallel in Britain where in peacetime the Army was broken up by the regimental system into many small fragments each protecting its own narrow interests. The Royal Tank Corps was prone to the same narrow-mindedness as the infantry regiments and many of its leading lights certainly did not emphasize the need for the kind of intimate co-operation with other arms that Guderian is discussing here. Harris and Toase, op. cit., pp. 27-51.
4. Although, by the outbreak of war, the Germans were beginning to introduce half-tracks for the infantry of the panzer divisions they still had very few of them in 1940. See M. Cooper, *The German Army* (Macmillan), 1978 p. 210. The motorized infantry, so vital to panzer division operations, were still largely mounted on motor-cycles and side-cars or in trucks. Most 'panzer grenadiers', as these infantry soldiers became known, continued to be truck-mounted until the end of the war. Naturally this meant that their mobility was more restricted than that of the tanks. Guderian did not, however, allow this fact to irritate him to the point of trying virtually to dispense with infantry altogether in mobile operations – the reaction of some of the tank enthusiasts in Britain. See Winton, *To Change An Army* (Brassey's), 1988, p. 110.

LIFE IN THE PANZER FORCES

After the rough outlines for the tactical and technical development of our new panzer forces had been worked out, we had to provide for their maintenance and training.

The first priority was to determine the establishment – which was more difficult than it might seem, when one considers that we could draw on virtually no experience to evaluate the needs of our future armoured forces. We reviewed the relevant passage in the history of the British and French in the last war, and we identified the kinds of demands that a future war might make on us. From all of this we deduced the following principles:

The panzer forces had to be ready to spring to action in the event of hostilities. It followed that their peacetime establishment must permit them to take the field without having to recall reserves on a large scale, or resort to untrained recruits. The following organization was the result:

Combat Company:
Company headquarters – the permanent entourage of the company commander;
Reconnaissance and signals personnel;
Double crews for the tanks;
Tank maintenance technicians;
Armourers;
Supply and internal service personnel.

Detachment Headquarters:
Reconnaissance platoon;
Signals platoon;
Medical officer;
Detachment engineer;
Workshop with supervisors and technicians;
Armourer.

Regiment:
Reconnaissance platoon;
Signals platoon;
Regimental band;
Regimental engineer.

On this basis we went ahead with choosing the garrison locations, drawing up plans for the barracks, and procuring training areas, ranges, clothing, equipment and weapons. While all due economy had to be observed, we made up our minds to do everything possible to promote the training of the forces and give them agreeable conditions of life.

Our choice of garrison locations was influenced strongly by the facilities available for training, and especially our requirement for extensive and varied terrain for exercises. On occasion a number of neighbouring garrisons were assigned a large common training area.

The barracks comprised two types of building. The accommodation blocks, together with rooms for offices and other administrative purposes, kitchens and canteens. Second, the technical accommodation with garages, workshops, tank parks, small-arms ranges and zeroing-in ranges. All the buildings were constructed to the same standards for accommodation and hygiene that obtained throughout the army.

Such are the surroundings in which the recruits for the panzer forces are trained. As elsewhere in the army they begin with basic instruction in posture, saluting, drill and weapons. The recruit intakes are admitted every October, and before long they are grouped according to their progress and aptitude as drivers, gunners or signallers, and receive specialized training as well as the basic groundwork. Only a matter of months later the men proceed to crew training; great attention is paid here to perfecting co-operation between the gunner and the driver, on which so much depends in action, and finally the crew fuses into a single unit where the individual identifies himself with the whole. Meanwhile the dispatch riders and scouts, the technicians and armourers all proceed with their specialized instruction. Needless to say the further training is not rigidly compartmentalized. The drivers and gunners have to learn enough of one another's skills to be able to help one another out, and understand what their comrades are doing. In addition many of the tank gunners are trained also as signallers.

The tank drivers are responsible for the condition of their vehicles, and they have to be able to carry out minor repairs by themselves or with the assistance of the other members of the crew. They must be skilled enough drivers to save the vehicle and the crew from needless wear and tear, and in action they must help the gunner by driving smoothly and making intelligent use of the ground. A high degree of vigilance is required to overcome the restricted vision provided by slits or the driver's optics, not least in conditions of heat and dust, cold and ice, darkness or fog. The drivers begin their instruction in open-topped training vehicles and then proceed to enclosed tanks. There is likewise a progressive increase in the demands that are made on drivers in overcoming difficult terrain and driving in formation.

The responsibility of the gunners extends to the weapons, ammunition, signals equipment and, in two-man tanks, also to command of the whole vehicle. In the semi-darkness of the enclosed turret and while being tossed about, the gunner must be able to display a consistent mastery of his weapon – clearing stoppages, and never failing to keep a close eye on the ground outside. The survival of the vehicle and its crew comes down to the marksmanship, courage and resolution of the gunner.

Live-fire training begins in the open air. It is continued in the tanks, first of all at the halt, and then with the tank moving at various speeds and in various directions – straight at the enemy, or with the hull at oblique or perpendicular angles to the alignment of the turret, and the gunner aiming variously at static or moving targets. The process is completed by battle firing at unit level. The use of simulators, such as firing from small-calibre weapons mounted in static mock-ups, spares the wear on vehicles and armament, and promotes something that is absolutely essential – constant practice in gunlaying.

The tank commander co-ordinates the efforts of his crew, and sees that his tank works with the rest of the unit. He frequently operates the radio and signals equipment. In turn one of the chief tasks of the company commanders is to ensure that the tank commanders receive the specialized training they require.

Crew training culminates in the process of unit training, which builds up to large-scale exercises and manoeuvres which are held towards the end of the military year.

Service in the panzer forces is fine and varied, and every tankman is proud to belong to this new arm, which is dedicated to the offensive. But panzer service is also demanding; it requires young men of sound consitution and mind, with cheerful hearts and determined will. Tank service builds small unit cohesion in a quite remarkable way; there can be no distinctions – officers, NCOs and men alike share the same testing conditions of combat, and everyone must play his part to the full.

Armoured equipment is expensive and rather complicated, which calls for a fairly large establishment of long-service soldiers. Just as the officers and NCOs need sound tactical and technical training, so a core of good engineers and technicians is required to attend and service the machines.

The Mechanized Forces School (*Kraftfahrtruppenschule*) provides the necessary knowledge and skills. It contains the headquarters and directing staff, and has instructional and experimental departments as well as running the tactical, technical and gunnery courses.

The tactical courses provide training and further education for the officers and senior ensigns of the mechanized forces, together with the training of reserve officers as company commanders, and a grounding for officers of the other arms in the principles and work of the mechanized forces.

The technical courses embrace the schooling of the NCOs, motor transport sergeants and maintenance technicians, the instruction and examination of specialized tradesmen, and the technical training of the officers and engineers of the mechanized forces.

The gunnery courses on the firing range serve both to train gunnery instructors, and to try out new gunnery tactics, equipment and instructional aids.

The instructional department provides the demonstration troops for the courses, and simultaneously trains its soldiers as NCOs.

The experimental department tests vehicles and components for the army as a whole. Its most significant commissions in recent times have been to test the durability of tyres of synthetic rubber (*Buna*) in extended service, and see how synthetic fuels perform over long journeys. The department also contains the 'sports staff' (*Sportstaffel*) which represents the Wehrmacht in major motor sports events.

The school has a splendid and thoroughly serviceable home at Wünsdorf, near Berlin.

This short review of our life and work ought to be enough to show that our panzer forces are engaged in a great variety of activities, and that they are in a state of active development. Almost every day we have to confront new problems, explore new avenues, and all the time we continue to make progress. We have room only for lively, open-minded spirits. We have to overcome the inertia of the individual, just like the immovability of the broad mass. Only when the panzer forces are full of verve, only when they are fanatically committed to progress, will they win through and achieve their great aim, which is to restore the offensive power of the army.

The Tactics of the Panzer Forces and Their Co-operation with the Other Arms

1. The Tactics of the Panzer Forces

We have seen in earlier chapters why and how tanks came into being, how they developed during the war and afterwards, and what were the thoughts of the godparents when the German armoured forces were born. We now turn from the history of events to the theoretical base. Here we will present a picture of the composition and tactics of modern tank forces, without losing our grasp on technical realities. It is our belief that the panzer forces have something of value to contribute to the army, and we will show how they fit in with the framework as a whole, and how they can best co-operate with the other arms.

By way of illustration we shall set our panzer forces the task of gaining a decisive victory. This they are supposed to accomplish by launching a concentrated surprise attack against a line of enemy field fortifications, aiming at a point which has been selected by our commander and which is favourable for the deployment of tanks. We have chosen the breakthrough of an enemy position as our example, in preference to alternatives such as mobile operations, envelopments or pursuits, since a breakthrough is perhaps the most demanding mission that could be set.

In this instance we are uncertain whether the defenders have laid out minefields. But we know for sure that their anti-tank weapons are capable of piercing our tanks at entry angles in excess of sixty degrees and ranges up to six hundred metres, and that they have about as many tanks as we have.

The attackers now have to choose their method of assault, considering first of all what elements of the enemy defence are respectively the most and the least dangerous. If their are mines in front of their positions, they could exact a heavy toll on our tanks.[1] Mines are therefore to be regarded as an enemy of an extremely dangerous order, and we must clear them at least in part before the armoured assault proper can break into the infantry combat zone. The work of identifying and clearing mines falls – as does the creation of practicable lanes through other obstructions – to the engineers. They approach the obstacles under cover of darkness or fog and, protected by artillery and machine-gun fire, set about clearing lanes for the tanks. Perhaps they will even have armour protection of their own; a number of foreign countries have for some years been conducting reasonably successful experiments with mine plough tanks and bridge-laying tanks.

All of this means that the panzer engineers will in all probability have to accompany the first of the assaulting waves in any breakthrough battle. They must be trained to locate mines and other obstacles under cover of darkness and mist and render them harmless, and they must be given the vehicles and equipment they need for this task.

We next have to reckon with the anti-tank guns.[2] These will be deployed throughout the entire depth of the zone of defence; those in the infantry combat zone will be already in position and ready to open fire, and farther to the rear some of them at least will be held ready to move. As regards their performance, we have already assumed that they can pierce our armour at sufficiently vertical angles up to a range of six hundred metres.

The attacker must now do something to attenuate the fire of the guns in question. He cannot afford to go on to assault secondary objectives as long as he is under their muzzles, which leaves him with no alternatives but to destroy them outright, or arrange to have them silenced or blinded by other weapons.

To destroy anti-tank guns our tanks must either take them under direct fire at the halt from behind cover, or overwhelm them by a mass attack. In addition the anti-tank guns can be suppressed by artillery or machine-gun fire, or blinded by smoke – this process can extend to guns lying outside the reach of the tanks themselves, for example in woods or villages that are not being subjected to direct attack, or in ground that is inaccessible to armour. But within the tanks' own combat zone nothing short of the destruction of the defence will do, if we are to develop the attack into a successful breakthrough. The best times for such an assault are at first light or when there is a light mist, for the defensive weapons cannot be used out to their full range, and the gunners will find themselves at a severe disadvantage when the tanks suddenly burst upon them. At the beginning of the attack the alarm will have gone out to the rearmost anti-tank units, and they will now be seeking to get into position. The attacking forces must therefore penetrate the defensive zone in great force and at great speed, so as to catch the anti-tank units while they are still moving up, and destroy them. Otherwise, when the sun comes up, the attackers will be suddenly confronted by a new defensive line extending immediately behind the forward battle zone, and it will be a costly and time-consuming business to break through, especially if the attackers have progressed beyond the reach of their own guns.

The enemy artillery batteries will be playing an active part in the defence, and they must be brought under attack at the same time as we are dealing with the anti-tank guns deep inside the defensive zone.

In our model we have postulated the existence of comparable forces of enemy tanks. But when exactly will they appear? They may no longer be able to render direct help to their own infantry, but they will certainly not be happy to see their artillery fall into our hands. It is when we take up the combat with the enemy artillery that we can reckon on the arrival of the enemy tanks. Many advantages are now on the side of the defenders – especially knowledge of the terrain, and the fact that they are in good order whereas the attackers will already be in some disarray. The tank's most dangerous enemy is another tank.[3] If we are unable to defeat the enemy armour the breakthrough has as good as failed, for our infantry and artillery will be unable to make any further progress. Everything comes down to delaying the intervention of the enemy anti-tank reserves and tanks, and getting in fast and deep into the zone of the hostile command centres and reserves with our own effective tank forces – and by 'effective' we mean forces that are capable of waging a tank battle. The best way of delaying the intervention of reserves is through aircraft, and this is probably one of their most important contributions to the ground battle. Useful work can also be done by long-range artillery, as long as we can determine with reasonable certainty the enemy approach routes or assembly areas.

As we have seen, the breakthrough battle imposes some pretty tough demands on the tanks. Success is probably attainable only when the entire defensive system can be brought under attack at more or less the same time. When the attack begins the enemy hinterland must be subjected to vigilant aerial surveillance so as to identify the movement of enemy reserves and direct our combat aircraft against them. The air forces must bend their efforts to preventing or at least delaying the flow of those reserves to the location of the breakthrough. On the ground the main weapons of the breakthrough remain the tanks. They will have overcome the minefields and the obstacles, and now they are deployed in depth in several echelons to attack and beat down the elements of the defence in rapid succession – the assembly areas of the enemy reserves, the command centres, the zone of the artillery batteries and the mobile anti-tank defence, and finally the infantry battle zone; of all these actions the victory over the anti-tank defences and the tank reserves is the most significant. If we win that combat, we at once have forces that are free to institute the pursuit and roll up those sectors of the front that are still holding out; the work of dealing with the enemy batteries and completing the clearing-out of the infantry battle zone may be left to fairly weak tank units, while our own infantry can proceed with exploiting the success of our armour. If, on the contrary, we fail to beat down the enemy tank-defences and defeat the enemy tanks, the breakthrough has failed, even if we manage to wreak some destruction in the infantry battle zone. In that case the battle will end as it usually ended in the World War, with a bloody and costly breakthrough that often left the attacker in a worse tactical posture – in a salient with vulnerable flanks – than before.

It is therefore of great importance to strive to bring the entire depth of the enemy defence under simultaneous attack.[4] This ambitious task can be fulfilled only by a large force of tanks deployed in sufficient depth, and with tank units and tank leaders who have learned to fight in large formations and, when the enemy puts up an unexpected resistance, to smash that resistance with speed and resolution.

In addition to depth, the breakthrough attack also needs a broad enough frontage to make it difficult for the enemy to bring the central axis under flanking fire. If a tank attack is so narrow that the area of the assault is actually enfiladed by machine-gun fire, the other arms will be unable to follow the tanks, and no durable success will be achieved.

We may summarize the requirements for a decisive tank attack by the concepts of: **suitable terrain, surprise and mass attack in the necessary breadth and depth.**

We will now describe in detail the tactics whereby the tanks fulfil this mission, and we will then be in a position to see how the other arms may contribute to the success of the attack.

We have assumed that most of our tanks are equipped with machine-guns and armour-piercing weapons, and are equally capable of doing battle with enemy tanks, anti-tank weapons and infantry. Our companies also incorporate a number of machine-gun tanks for reconnaissance, communications and the less demanding combat tasks. The tank detachments should be equipped with tanks with big guns on the model of the British 'close support tanks'. Several detachments form a regiment, and several regiments form a brigade.

Tank forces are directed by radio, and the smaller units from company downwards also by visual signals. As long as radio silence has to be observed, orders and reports can be transmitted by means of aircraft, vehicles or telephone. Commanders ride in the command tanks, which are followed by the necessary radio tanks for secure communication with superiors and subordinates. We should strive for aerial reconnaissance of the area of the attack beforehand. The average marching speed of tank units is 20 kilometres per hour in daylight, and 12-16 kilometres at night; in favourable weather and on good going the combat speed is 16 kilometres per hour.

The assault is preceded by **reconnaissance** covering the approach routes and assembly areas, the likely ground for our attack and the enemy we may expect there. The basis for the plan of attack is provided by the study of maps, evaluation of aerial photographs, interrogation of prisoners and other sources of intelligence.

Successful surprise is of absolutely fundamental importance for the chances of victory. The greater operational and tactical mobility of armoured forces lends itself particularly well to surprise, provided that the preparations for the attack on the part of all arms are kept as concentrated and as short as possible, the **approach marches** are accomplished under cover of darkness, the flow of supplies is concealed, and the movement of traffic at night is kept under close control. We assume that the tank forces are brought up to their assembly areas by night and without lights, using routes that have been identified beforehand, and which are well signposted and have been kept clear for them. As a general rule the assembly areas should be out of range of the enemy artillery, so that our forces can make themselves ready for the battle, change their crews after the long approach marches, distribute rations and establish contact with the other arms. However, departures from this rule may be occasioned by difficulties of terrain and other considerations.

The forces make their **initial deployment** (*Entfaltung*) for the attack from their assembly areas. This must be accomplished early enough to enable them to cross their own front line at the appointed time. Initial deployment means taking up formation in the breadth and depth intended for the later attack, though the individual units mostly remain in column of march in order to exploit the available routes, make an easier passage of defiles, and pass with the minimum of disruption through the other forces already in place, not least the signals elements. We will use feint attacks, smoke-screens, artillery fire and air activity to keep the enemy in the dark as to where we intended to make our break-in.

Immediately before going into action, usually in the last available cover, the forces accomplish their **full deployment** (*Aufmarsch*) into combat formation; this is a particularly difficult manoeuvre if the terrain has restricted the initial deployment to an excessively narrow frontage. Our tank forces must be trained to carry out speedy approach marches and deployments by night, or they will find themselves engaged in lengthy preparations under the noses of the enemy, liable to detection and heavy casualties, and they will jeopardize the element of surprise.

From the deployment to the opening of the **fire-fight** the attackers make whatever use they can of the ground and then, when the enemy is in sight, they deliver the assault at full speed. For the fire-fight, however, the speed must be moderated and, if the situation permit, it can be conducted from the halt.

The outcome of the fire-fight essentially determines the outcome of the tank attack. The individual lines of the attack, and the leading wave in particular, must present a strong front with plenty of fire power; the following lines lend immediate support and keep the gaps constantly filled. It is easier for the anti-tank guns to pick off individual tanks when the attacker tries to break into the enemy battle zone in a thin, dispersed assault; when the breakthrough is made on a broad front and accompanied by heavy fire from the tanks, the defensive array is much more likely to be overrun, broken through and rolled up from the flanks and rear.

We now come to **combat formations**. These are of greater moment for the armour than for the other forces, because they must permit the tanks to use their weapons effectively without getting in one another's way, and because the formations must facilitate the use of ground and thereby enable the machines to take cover and lend mutual support. The simpler the formations are, the more easy they will be to maintain, and the speedier the transmission of orders will be.

The smallest combat unit is the platoon, which is made up of between three and five machines in the case of the heavy and medium companies, and five and seven within the light companies. Platoons are not usually subdivided further. In combat the platoons move across country in line or wedge with intervals of about fifty metres between the vehicles. The platoon commanders usually take up station in the middle or at the head of the array; they are responsible for maintaining the formation, speed and place of the platoon within the company, and they must also maintain reconnaissance or at least observation to the front and any open flanks and, in the case of the last line, to the rear as well. The march formation is the column, and on the battlefield the double column as well.

For the attack the companies are deployed in waves (*Wellen*), the detachments in **lines** (*Linien*), and the larger tank formations in lines of battle (*Treffen*). All the commanders stay well forward during the assault, so that they can keep their units constantly in view, and bring their personal influence to bear without delay.

Light companies in the first line are often assigned individual platoons of medium (gun) tanks for close support.

Each line of battle and each of the component units is to be given a clearly defined **combat mission**. The nature of the task will depend on the considerations already outlined. If it be feasible, the objectives of the attack and the relevant landmarks should be pointed out to the subordinate commanders on the spot, thus ensuring that the direction of the attack will be maintained despite any smoke or dust. If this prove impossible, because of irregular ground, mist or darkness, we resort to the compass.

The intervals and formations of the rearward lines and battle lines are defined by the nature of the mission, the terrain, the deployment of the next forces to follow through, and finally by the impact of enemy weapons and how the combat actually goes. Here it is important to remember that the rearmost units must be in a position to lend speedy support to those in front, while maintaining their own freedom of movement – this is the way to avoid bunching up and offering targets to hostile artillery and aircraft, if the movement happens to be interrupted. It will also permit the rearmost units to be diverted in other directions.

The attack moves at speed, and as soon as it closes to effective range we go over to the fire-fight. It will be waged by the individual lines of battle according to their particular missions. The most valuable characteristic of the tank is its capacity to deliver effective and close-range fire against clearly identified targets, and to destroy them with just a few rounds. The tank also has the capability of suppressing areas suspected of being held by the enemy, and objectives that may or may not be defended – though this work uses up a great deal of ammunition. We distinguish between fire at the halt and fire on the move. The first should be preferred if the combat situation and the cohesion of the unit permit the tanks to stop and fire; but there will be no alternative to firing on the move if that is dictated by enemy action or the necessity of keeping the attack together. Firing at the halt permits the gun to fire probably to good effect out to the limit of visbility; when firing on the move the machine-guns are effective up to 400 metres and the guns up to 1,000 metres.

It will often prove possible for the progress of the leading units to be covered by the rearmost waves and lines, which can lay down a long-range supporting fire at the halt.

In addition to their fire, the tanks can **crush** enemy equipment, obstacles and cover and, on occasion, enemy personnel. This capacity depends on the weight and engine power of the tank, and to a certain degree on its climbing ability and external shape.

The so-called moral effect of the tank depends ultimately on its physical effect – the impact of its fire and its crushing power. In the last war the moral effect was very potent, despite efforts to diminish it, and the fundamental reason was that the Germans had insufficient anti-tank defences and hardly any tanks of their own. In future, the moral effect will be that much the less as an opponent attains greater equality in those respects. It is therefore of the utmost importance to make an accurate evaluation of the enemy capacity in tanks and anti-tank weapons; here we must take as much account of the technical side – how the equipment performs – as the organizational and tactical aspects, which take in questions of leadership and deployment.

Tanks are naturally most effective when they are directed against completely defenceless targets or at least targets incapable of putting up an adequate defence. Tanks are also effective against badly concealed objectives, or those positioned in terrain that is accessible to armour; conversely the effect falls away in the face of strong opposition, good concealment or cover, and objectives in restrictive or prohibitive ground.

In this context we must add a few words on **tank versus tank combat**. Military literature tends to steer clear of this subject, invoking as an excuse our lack of experience. This attitude cannot be sustained over the long term. We will be unfailingly presented with the reality of tank versus tank action in the future, as we have already established, and the outcome of the battle will depend on the issue of that combat, irrespective of whether we are cast in the role of attackers or defenders.

In the last war there were just two clashes of German and British tanks, at Villers-Bretonneux on 24 April 1918, and at Niergnies–Séranvillers on 18 October of the same year. We shall present the essential features in outline.

2. THE TANK ACTION AT VILLERS-BRETONNEUX

(See Sketch Map 14)

At 0345 on 24 April 1918 the German artillery opened up in preparation for an attack on the sectors held by British III Corps and French XXXI Corps. The bombardment was sustained with great violence for three hours and at 0645, in dense fog, the German attack was launched on a frontage extending from north of Villers-Bretonneux to the wood of Sénecat (three kilometres south-west of Thennes). There were three divisions in the German first line, namely 228th Jägers, 4th Guards and 77th Jäger Reserve. They took under command the following numbers of tanks: three with 228th Jäger Division; six with 4th Guards Division; four with 77th Jäger Division.

These thirteen machines were all we had available at the time.

When the bombardment began the German tanks went off to the assembly areas, from where they set out a few minutes before the attack opened, in order to be able to cross their own front line on time. At first the assault made slow progress, since the thick fog limited visibility to fifty metres, and the contact between tanks and infantry was immediately lost. Whenever the British offered the least resistance the German troops stopped, hung about and sometimes even fell back. Towards 1100, however, the fog lifted, and the infantry resumed contact with the tanks and began to make more rapid progress.

The three tanks with 228th Jäger Division gained their objective, and were then ordered to assemble at Viencourt.

On the frontage of the central division, 4th Guards, four of the tanks likewise reached their objective; one tank stuck in a shell crater, and the sixth broke down from engine failure.

On the sector of the left-hand division, 77th Reserve, one of the tanks managed to silence several machine-gun positions and stretches of trench, but towards 0845 it became lodged sideways in a sandpit and was unable to get out (it was later retrieved and spirited away by a French recovery

THE ACTION AT VILLERS-BRETONNEUX
Tank against Tank, 24 April 1918.

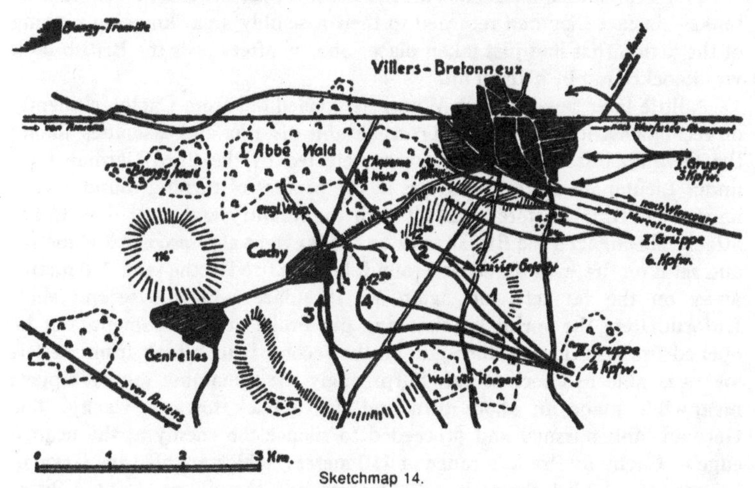

Sketchmap 14.

team working between the lines). The second machine destroyed a number of machine-gun positions and actually came within 700 metres of the edge of Cachy village, which it took under gun and machine-gun fire.

The third tank likewise accounted for a number of machine-guns, cleared several hundred metres of trench and, having reached its objective, was on the point of turning around to return to its assembly area. The fourth tank was doing the same, after it had joined in the fire attack on Cachy.

At this juncture the second tank spotted three British tanks emerging from the southern tip of the wood of Arquenne – two females in the lead and a male Mark IV following. What had been happening? As a safeguard against any German attack the British had sent a tank company ahead into the wood of Blangy from where a platoon of three tanks was pushed first into the wood of l'Abbé and then, because of the German artillery fire, to a position behind the southern tip of the Arquenne wood. Between 1000 and 1100 the platoon in question received the rather vague order to secure the Cagny blocking position.

As soon as they left cover the two female Mark IVs spotted four German tanks and at once advanced to attack the nearest machine. The German tank immediately faced about and, taking advantage of some cover, opened fire. One of the British females was seen to be hit, and both machines gave up the attack. The third British tank, the Male Mark IV, now materialized at two hundred metres from the German tank and scored several hits with a 57mm gun. Five of the crew were killed; the survivors abandoned the tank, though they were able to reoccupy it later and bring it back behind their own lines.

The commander of the male Mark IV lost contact with the other German tanks – in fact they had returned to their assembly area, knowing nothing of the action that had just taken place. Shortly afterwards the British tank was knocked out by a direct hit.

A little later seven British Whippets pushed out from Cachy, evidently on reconnaissance. They caused considerable disorder and casualties among the German infantry, but they were detected by the third German tank under Lieutenant Bitter, who was in the process of turning round, as we have seen. Bitter restored contact with the infantry and went over to the attack. He engaged the British tank on the far right at a range of 200 metres and set it on fire with the second round; he then fired at the tank 700 metres away on the far left and again set it ablaze with his second shot. Unfortunately the spring of his firing pin broke at this moment, but he opened fire with his machine-gun on the second British tank from the left and was able to knock it out. Surprisingly the remaining four Whippets meanwhile made an about turn and drove back towards Cachy. The German tank pursued and proceeded to silence the enemy at the nearest edge of Cachy by fire at a range of 150 metres, which enabled the German infantry to establish themselves 400 metres from the village. At 1445 Bitter returned to the assembly area.[5]

3. THE TANK ACTION AT NIERGNIES–SÉRANVILLERS

On 8 October 1918 the British attacked between Cambrai and Saint-Quentin with the support of sixteen battalions of tanks, of which 12th Battalion went into action south of Cambrai, where it was distributed among four of the assaulting corps. Initially the attack went well, but in the morning mist it ran into a counter-attack by German tanks – namely ten captured British Mark IVs, which were advancing under cover of a smoke-screen. The leading British tank commander took the approaching black shapes for friendly machines, as was only natural, but he was disabused when he came under fire at fifty metres. The British were able to hit the leading German tank, but four of their own spearhead tanks were rapidly put out of action, and it seems that some at least of the crews were unaware of the enemy presence as they hastened towards their doom. A British tank officer managed to get hold of a gun which had been captured from German troops, and he was able to put a further German tank out of action, leaving the Germans with two machine-gun tanks. One of the surviving machines was disabled shortly afterwards, while the other had to make off to escape the attentions of a British machine.

North of Séranvillers, meanwhile, two British gun tanks encountered two German machine-gun tanks and, inevitably, soon knocked them out. In this way the British were able beat off the German counter-attack. Their own infantry had fled before the German tanks, but now they came on again and seized their objectives.

These were the only tank actions of the war. They were admittedly small in scale, but they furnish some valuable lessons:

1. When tanks advance against an enemy who has weapons that can penetrate their armour, while they cannot repay them in the same coin, they have no alternative but to retreat. More specifically, tanks armed with machine-guns only are powerless against tanks that have guns as their primary armament, and possess armour that is impervious to armour-piercing rounds of small-arms calibre. This old truth has been confirmed recently in Spain.
2. The tank's most dangerous enemy is another tank. As soon as a tank force identifies enemy machines, and is in a state to do battle with them, that force is duty bound to drop all its other missions and engage in combat. This also happens to be the best service we can render to our own infantry, since they will be in as much danger as our tanks if the enemy manage to break through with an armoured counter-attack.
3. Tank versus tank combat is decided by fire. We must therefore bring our tanks to within effective range of the enemy, making use of ground so as to present a small and elusive target. The tanks must strive to improve their chances of a hit by firing at the halt, especially when they first open fire.[6] It is also important to make use of good light and favourable wind.
4. We cannot be content with training for individual combats of tank against tank. We must reckon on the appearance of large forces of tanks, and it is

much more useful to work out how to manage combat on this scale. In actions of this kind we cannot avoid having to fire on the move:
(a) as a protection against the increasingly effective enemy fire;
(b) in order to frustrate enemy outflanking or enveloping movements, whether through committing reserves or manoeuvring with the forces that are already engaged;
(c) to outflank or envelop the enemy through our own reserves, and concentrate our forces so as to pit a greater number of our own tanks against a smaller number of the enemy. It follows that armoured combat demands strict discipline, good fire control and good training on the ranges. Command, and especially the commitment of the reserves, is facilitated by maintaining formation among the units, and observing the prescribed speeds. Supposing the armament of the combatants to be approximately equal, victory in the tank battle, as in the combat by other arms, will go to the side that has the advantages of superior, firm and resolute command, and has laid the foundations for such leadership in good time.
5. The other weapons, and the artillery and anti-tank guns in particular, must not be content with the role of passive spectators during the armoured battle. On the contrary, it is their obligation to contribute to the utmost of their power to the victory of our tanks. Here again the lessons of Spain have confirmed those of 1918.

The tank attack terminates with the **redeployment of the units** for further missions – these may consist in completing the breakthrough, pressing ahead with the pursuit, rolling up sectors of the front that are still holding out, or checking or smashing enemy reserves that are on their way up. In the event of our attack failing, we must reassemble our forces in a suitable location.

Assembly areas can seldom be fixed in advance, and their location is usually determined only after the attack has come to an end. Assembly areas must have cover against direct fire and aerial observation, must facilitate readiness of combat at short notice, and must be properly secured. The assembly areas are where ammunition and fuel are replenished, the men are fed, losses are replaced, and burnt-out units replaced by fresh ones. If necessary, the combat supply train must be brought up to the same locations.

4. THE CO-OPERATION OF TANKS WITH THE OTHER ARMS

Tanks are unable by themselves to meet all the combat tasks which come their way; the other arms will be needed as well, for example to deal with difficult terrain, artificial obstacles, or anti-tank weapons sited in 'tank-prohibited' ground. In this requirement the tanks differ in no respect from the other arms, and inter-arm co-operation is therefore a matter of fundamental importance. Everyone is agreed on this basic point, but the difficulties creep in when one goes a little farther, namely when we begin to ask about the 'how' of co-operation.

Here we may distinguish between three schools of thought. One of them continues to regard the infantry as the 'Queen of the Battlefield', just as in earlier days; the single great weapon which all the other arms must dedicate themselves to serve, if necessary by denying themselves some important advantages. For such people the infantry remain the 'Bearer of Victory'. They focus their attention on the fire of the enemy heavy weapons, which represents the immediate threat to the foot soldiers. The main function of the tank is therefore seen as the destruction of those weapons, and the tanks are supposed to accompany the infantry not just at the start of the battle but right through to the end. These folk forget that clearing the infantry battle zone is not such a mighty undertaking after all, and that it may be accomplished safely and easily by a handful of machine-gun tanks – as long as those tanks are given the opportunity to go about their work undisturbed. This opportunity was a reality in 1918, but it certainly does not exist nowadays. In fact the defence has acquired such tremendous power that the enemy anti-tank guns and artillery observation posts must be eliminated before the assault begins, otherwise the tanks that are sent into the infantry battle zone will be unfailingly destroyed. From the viewpoint of the tank forces, the most important mission is not infantry support, but to destroy the enemy anti-tank defences and suppress or blind the enemy artillery; when that has been accomplished the tanks will indeed render the infantry the support they need – rapidly, thoroughly, exhaustively and economically.

The same holds true from the perspective of the high command, if we are striving for decisive victory on the grand scale, and not just an infantry attack which has limited objectives and which is carried out at the infantry pace. It makes no sense to dispatch tanks straight into the infantry battle zone to seek out hidden machine-gun positions, while immediately to the rear the defenders are left in peace to build a new line or prepare counter-attacks. These tactics were tried dozens of times during the World War – they failed on every occasion, and they will inevitably do so in the future. A modern style of leadership, in full command of the weapons at its disposal, will instead aim for a rapid decision, and will therefore place demands of an altogether higher order on its tanks – demands which must stretch them to the limits of their capabilities, since the commander will otherwise throw away his trump card.

The proper assessment of the limits of those capabilities is a matter of decisive importance. One school, as we have seen, wishes to draw the bounds too tightly, but another is inclined to extend them altogether too far. Its adherents dream of grand operations, raids into the enemy rear, surprises, and fortresses and fortified zones which will fall into our hands with effortless ease. But it is more than doubtful whether a future war will ever begin with the freedom of movement which the Germans enjoyed to a certain degree in 1914, after they had captured Liège. In all likelihood there will be a need for an immediate battle involving fortresses or defended positions; the attacker must achieve a breakthrough if he is to win freedom of movement, and even then he must exploit very smartly if the front is not to congeal almost immediately; for this is just the stage when the defender has a chance to use his own mobile forces to telling advantage.

This chain of argument has given rise to a third school of thought. Its protagonists argue that, without departing from what is technically feasible, tanks can do something more than act as mere train-bearers to the infantry, or leaders of the infantry cortège – the most that our first school would allow. At the same time the third school takes all due account of the actual obstacle-crossing capabilities of tanks, and their real prospects in combat against anti-tank defences and the enemy armour – they have no wish to throw their tanks away to no purpose. This school of thought also bases its case on a number of further principles:

– that infantry are strong on the defensive, but their offensive capacity is feeble, or at least slow-acting, because of the defensive power of the modern infantry weapons they have to face;

– that artillery fire of the most powerful kind is still insufficient to guarantee rapid and deep break-ins of the enemy battle zone;

– that the existence of enemy motorized and armoured reserves rules out the possibility of decisive victory through conventional breakthrough tactics.

To this way of thinking there is a need to gain a tactical decision with the speed consonant with the modern era of aircraft and tanks, and then exploit it into the operational dimension. How do we do it? – by elevating the air and tank forces to the status of main arms of the service. Only experience of actual combat will show whether these efforts will be rewarded. One thing is certain, that the former offensive tactics, using conventional weapons, failed to achieve a decisive success in the course of four years of bloody war. In other words they have no future validity.

We therefore aim at achieving a sweeping victory by the means advocated by our third school, namely a breakthrough which will lead to a pursuit, and the rolling up of the intact sectors of the enemy front. When co-operation with the other arms is discussed, it must therefore be in the context of the requirements of the armoured attack.

For this purpose we have derived from the last war the three requisites for the success of a tank attack: suitable terrain, surprise, and the concentration of all available forces at the decisive point – in other words attacking *en masse*. The frontage must be broad enough to prevent the axis coming under flanking fire; otherwise the tank attack itself may be successful, but the unarmoured forces, especially the infantry, will be unable to follow up. In the last war the French and British attacked on frontages which already extended to twenty and thirty kilometres; in the war of tomorrow the attacks will be at least as broad, and very likely considerably deeper, because of the strength of the defences which we have to overcome, the greater distances to the final objectives, and the need to roll up the stretches of the front where the enemy are still holding.

We have no desire to impose a rigid tactical framework,[7] but in general we foresee the assaulting armoured forces deployed in four lines of battle: the first line should pin down the enemy reserves, including the tanks, and knock out the headquarters and control centres; it should eliminate the anti-tank weapons it finds on the way there, but not otherwise get embroiled in fighting. The second line has the task of destroying the enemy artillery and any anti-tank defences active in its neighbourhood. The third line should bring our own infantry up through the enemy infantry battle zone, and in the process eliminate opposition from the infantry so thoroughly that the supporting arms of the tanks are able to follow up. We can form a fourth line only if we have a very considerable number of tanks, but it serves as a reserve at the disposal of the high command, and for rolling up intact sectors of the front. The whole, mighty assault must break into the enemy defences simultaneously and on a broad front, and press on to its objective in a series of continuous waves. After they have achieved their initial tasks it is the common responsibility of all the lines of battle to press forward, so as to be at hand for the imminent tank battle. Our first line must be very strong, for the armoured battle is difficult work and will invariably fall to its lot; the second and third lines may be weaker. The strength of the fourth line depends on the situation and the terrain. If we find firm support for the flanks, we can probably get away with covering them by anti-tank guns and other weapons; open flanks and wings usually demand protection through several echelons of tanks deployed in depth.

The attack is preceded by the process of reconnaissance and scouting, the approach and the assembly.

Reconnaissance is primarily the work of aircraft, supplemented by motorized reconnaissance detachments or other units already in contact with the enemy. The agents of reconnaissance must be faster than most at least of the troops who are coming up behind them, and they must transmit their reports to their directing officer with the least delay. Reconnaissance for a tank attack must establish the enemy's defensive deployment, especially the arrangement of his reserves, and among these again the anti-tank guns and the armoured forces in particular. Reconnaissance must extend well behind the front line, since motorized forces can cover considerable distances in a matter of hours. It is on the basis of this intelligence that the commander determines not only the missions and the offensive deployment of his armoured forces, but the intervention of the air forces in the ground battle. Reconnaissance can further identify the natural and artificial obstacles, and photographic reconnaissance is especially useful in this respect. The findings of aerial reconnaissance can be supplemented by ground reconnaissance and scouting. Careful study of maps is essential, if one wishes to avert unpleasant surprises.

It is most important that the process of reconnaissance and scouting should not betray the location of the intended attack to the enemy; security is also to be observed with regard to our own troops. Before the Battle of Cambrai, for example, General Elles and Lieutenant-Colonel Fuller went to the length of disguising themselves by removing their patches and wearing dark glasses.

Concealment is of fundamental importance for the **approach** and **assembly** of our forces, if we are to maintain surprise. There was once a time when no great importance was attached to concealment, but the events of 1917 and 1918 speak so clearly to the contrary that we must enter here into some further detail. Concealment against aerial reconnaissance is effected through speed of assembly before the attack opens, by moving up in darkness without the use of lights, and by sedulous concealment of the assembly areas. Radio interception is frustrated by observing strict radio silence until the battle opens. Anti-aircraft defences and measures against aerial reconnaissance must be thought through carefully, but should not be designed in such a way that they might of themselves give a clue as to our intentions.

Once the attack begins, the operational and tactical aerial reconnaissance must be supplemented by **combat reconnaissance**. Its findings are invaluable to the tank commanders, and they must be got to them as rapidly as possible, for example by radio, or throwing messages from aircraft – minutes can be of decisive importance when the enemy have established new defences or even brough their own tanks on to the scene. Smooth co-operation with the air forces is the product of frequent joint training.

With the opening of the attack two further weapons acquire considerable significance for the tank forces, namely the artillery and the engineers.

The immediate issues concerning the **artillery** are whether the tank attack is to be ushered in by a bombardment of longer or shorter duration, and whether we will dispense with this particular calling-card altogether. Opinions on the point are divided. Some officers maintain that fire is indispensible, and that 'resort to artillery firepower is an essential preparation for the armoured attack'. The opposing school cites the examples of Cambrai, Soissons and Amiens and wishes to delay artillery support until the assault begins.

One thing is indisputable – the shorter the artillery preparation the better. A prolonged bombardment betrays the location – and to some degree also the time – of the attack, and permits the defenders to place their reserves in readiness, occupy rearward positions, and perhaps also prepare withdrawals and counter-attacks which will occur at unexpected and unpleasant places – as at Reims on 15 July, with the riposte following on 18 July at Soissons. Long bombardments churn up the ground and make it difficult going for all arms, but particularly inconvenient for the tanks, which are supposed to get ahead quickly. However a short bombardment may be indispensible to provide protection for the engineers, if they have to open lanes for the tanks by clearing obstacles, or building pathways over watercourses or swamps.

Assembling large quantities of guns and their ammunition is a time-consuming and all too visible process which may put surprise at risk. It seems best to do away with the artillery preparation altogether, and burst upon the enemy with the advantage of complete surprise. Once the attack has opened, however, it must be supported by the artillery.

The missions of the artillery are to suppress targets and geographical features which the tanks cannot tackle themselves (villages, for example, or woodlands, steep hillsides, watery or boggy terrain), to hold down or blind likely observation posts and locations of anti-tank weapons, or to destroy identified targets which might impede the tanks. Long-range artillery can usefully seal off the area of the attack, and harass identified or suspected command centres and assembly areas, or be held in readiness to cover the tank attack as it progresses.

Once the tank attack opens, the artillery usually has to lift its fire from the area of the assault. When the artillery is in static positions it can support the armoured assault only to the limit of vision of the observers. If the observers are able to go along with the attack, the cover is extended to the full range of the guns, but when this limit too is reached the artillery will have to change position, which will detract from its effectiveness.

To follow a successful tank attack is absolutely out of the question for horse-drawn artillery, and quite difficult even for guns towed by vehicles. What do our tank forces want and need? It is artillery support that is fast-moving and sufficiently well-protected to follow immediately behind them. As well as having its specialized form of mobility, the supporting artillery requires specialized skills, and these are acquired through joint training with the tanks. Here it is worth mentioning that it is more demanding to direct armoured mobile artillery than the conventional artillery assigned to the infantry divisions. The mobile guns have to do their work over a shorter period of time, and have to deal with a greater variety of targets. The armoured attack has no need of a concentrated, pre-planned artillery barrage; no battering away at positions until they are ready to be stormed. Its call is for a responsive, fast-moving and accurate type of artillery, capable of following the assault with all the speed that it will acquire when it is commanded properly.

Among the other questions relating to the use of artillery in the armoured attack are the use of **smoke** and the application of **chemical weapons**.

If Nature has failed to oblige us with fog to blind the enemy defensive weapons and observers, the artillery must provide a substitute in the form of smoke. In accordance with the timetable of the tank attack, we will lay down smoke to blind enemy observation posts and the suspected locations of anti-tank weapons, and along the edges of villages and woods; the effect will extend over a certain length of time, and is intended to enable our tanks to approach unseen, or drive past unscathed on their way to execute an enveloping movement. As the attack progresses, smoke will also be put down at the gunners' own initiative or at the call of the tanks. Here the purpose will be to blind identified targets such as anti-tank guns or, if necessary, enemy tanks as well. If the attack turns out badly, smoke can facilitate our disengagement from the enemy.

In addition to smoke shells and generators, tanks have the capacity to create their own smoke. This is attended with certain disadvantages. As the source of the smoke is usually identifiable, this may betray the location of the tank, or the direction in which it is going. Tanks often end up driving in their own smoke and become blind, or they stand out all too clearly against their self-generated cloud. As a general rule, therefore, we advocate the use of smoke by tanks only in favourable wind conditions; however it can prove a useful aid in disengaging from the enemy, as already mentioned.

Chemical weapons have little effect on tank crews. Inside the tank, protection against poison gas is offered by masks and overpressure; the structure of the tank itself gives protection against blistering agents such as mustard gas (Yellow Cross). One of the great advantages of tanks is their imperviousness to chemical weapons.

So much for artillery. **Engineers** too must lend the tanks a hand during the preparations for the attack or, at the latest, at the beginning of the assault. Before the attack gets under way it is important to make the approach routes practicable, not least when they lie across country. Route signs are needed to guide movement at night. The engineers must also provide pathways over watercourses, swamps or soft going, and reinforce bridges that are too weak to bear tanks.[8]

The work of the engineers becomes far more testing when the attack begins. The defenders will have done their best to site their strongpoints in ground that is difficult or impossible for tanks, and where this is not available they will have covered their positions by obstacles, and mines in particular. Their identification and clearance is an extremely difficult but extremely urgent task that is mostly carried out immediately before the enemy front line – which means inside the enemy's most effective range. Moreover the work usually has to be done in the greatest haste, since the defenders take the beginning of obstacle clearance as the sign of an imminent attack, and they use every minute to consolidate themselves. Even when the work of the engineers goes forward under the protection of artillery and smoke and the heavy infantry weapons, there can be no guarantee that the defence has been suppressed; when engineers are assigned to work with tanks, there is therefore no alternative but to put some at least of their personnel in armoured vehicles fitted with equipment for detecting and clearing mines. Amphibious and bridge-laying tanks are good for surmounting watercourses, and we have seen prototypes of such vehicles in Britain, Italy and Soviet Russia. Speed of execution must be the overriding requirement for the engineers, just as for the other arms involved with working with the tanks. These *Panzerpioniere* therefore need special gear and training. In addition the army engineers will find themselves concerned not only with anti-tank defence, but co-operating with our own armour on the attack.

We will now assume that artillery support and skilful work by the engineers has helped the tanks to break into the enemy defensive zone, and that the attack is well under way. The break-in will occasion a falling-off in the enemy resistance in the outer combat zone, but a feverish activity further to the hostile rear, for all the available reserves will be streaming towards the fighting – in the air and on the ground – armoured and unarmoured forces alike. The immediate responsibility of halting the flow of enemy reserves falls to the **tactical aircraft**, which must cast aside all other work in favour of intervening in the ground battle at this decisive moment. However the slackening of resistance in the outer combat zone must be exploited by every means; every arm, not least the infantry, must press forward at all speed.

Before the tank attack opens our **infantry** will have been making preparations to support the armour and exploit its progress. Some of the heavy infantry weapons will keep the battlefield under surveillance, ready to shoot up anti-tank guns as they appear, and some will be participating in the general fire plan to maintain a suppressive fire on areas by-passed by the tanks. The teams of horses for the supporting weapons will have been kept as far forward as is consonant with safety, and likewise the reserves will have closed up tightly in expectation of moving forward into the attack.

As soon as the tanks have had any noticeable effect on the enemy, the opportunity must be exploited without delay; in some places the results will be transitory, and a number of the enemy machine-guns will re-open fire. The best recipe for a sure and economical success is therefore to exploit the initial surprise of the enemy by means of an immediate advance. The infantry should be under no illusions: the tanks can cripple the enemy and knock a hole in their defensive system, but they cannot dispense with the need for infantry combat.[9] This is by no means a bad thing for our own infantry, for it demonstrates that they have an essential part in the common battle.

The infantry combat will now focus on whatever strongpoints have been spared or left unidentified during the tank battle. The task of the infantry will be facilitated by the fact that the strongpoints can be turned and surrounded by exploiting the avenues already cleared by the tanks. Moreover some of the tanks are usually assigned for co-operation with the infantry, at least for the duration of the combat in the infantry battle zone.

We have every confidence that we will be able to render the infantry material help when the armoured attack itself turns out well. But we must emphasize yet again that the essential precondition is victory in a thrust which the tanks carry deep into the enemy defences, aiming at their principal foes – the hostile tanks, anti-tank guns and artillery.

The tanks will go ahead of the infantry when an extensive tract of open ground has to be traversed before the break-in. When the two sides are in close contact and the terrain favours the attack, the tanks will attack simultaneously with the infantry; the infantry will have to attack under artillery cover ahead of the tanks when we need to overcome initial obstacles – a stretch of river, for example, or barriers or minefields – before the tanks can intervene.

There is no need for the infantry and tank attacks to proceed on the same axes; of the two it is the armoured thrust that is the more strongly influenced by the topography. If, however, the axis is the same, and the tanks have no alternative but to drive through infantry already deployed, the infantry must take up formations that will permit their own advance to continue at speed, while enabling the tanks to distinguish the infantry units in half light or in fog. Otherwise there will be a danger of accidents and of our own troops opening fire on one another.

It is a physically taxing business for infantry to accompany a successful tank attack on foot; they must be specially conditioned for this purpose, and be fitted out with light equipment and suitable clothing. The quickest and most effective way to exploit the success of the tanks is by motorized infantry, especially if the soldiers' vehicles are armoured and have complete cross-country mobility, as with the French *Dragons portés*. If such rifle units were united permanently with tanks in a single formation, it would form a comradeship in arms in time of peace – a comradeship that would prove its worth when we come to seek the decision in the field. The benefits in terms of morale would be at least as great as the tactical ones.

We hear some people claim that infantry are powerless without tanks, and that every division of infantry must therefore take a detachment of tanks under its command. Others arrive at the same conclusion from precisely the opposite direction, maintaining that infantry remain the chief arm just as before. Whether they underestimate or overestimate the infantry, they nevertheless agree on one thing – the tanks must be split up! Whatever one may say about the present offensive capacity of infantry, it is undeniable that just about the worst service you could perform for the infantry would be to divide the armour, if only in part. Many infantry divisions will of necessity have to fight on the defensive for greater or lesser periods of time; they can make do with anti-tank weapons. Willy-nilly, other infantry divisions must go over to the attack, and many of them will do so over ground that is difficult or impossible for tanks. If we were to subordinate tanks to all of these divisions as organic elements, we would end up with that many fewer tanks at the point where we seek the main decision, and where their intervention would be the most rewarding. This is when the infantry really need tanks, and if they are deprived of them by some organizational blunder they will have to pay for it – as always – in their blood. A number of discerning officers of infantry are in full agreement on this point, and they urge that armour be concentrated in large formations.

We have already touched on the **co-operation between air forces and armour**. As already mentioned, aircraft can halt the movement of enemy reserves, especially the motorized and armoured forces, towards the location of the decisive battle. There may well be a need for air attacks on road and rail traffic, command centres and the whole communications system, as well as troop accommodation, and identified assembly areas, batteries and anti-tank forces. We do not underestimate the difficulty of hitting small, well-camouflaged targets, or of destroying moving objects when we have no clear idea where they will be when the offensive opens. However aircraft worked to telling effect as long ago as 1918, and the attacker can hardly dispense with their co-operation nowadays.

The missions just outlined can be accomplished still more thoroughly and permanently by **paratroops and air-landing forces**. Comparatively small airborne forces can hinder the intervention of unarmoured reserves in a very unpleasant way. Important locations in the enemy rear can be seized from the air, and turned into strongpoints and logistic bases for the approaching tank offensive. In co-operation with tanks, the airborne forces could inflict considerable damage and disruption on rearward communications and establishments, and even take airfields under attack. At the very least airborne forces will provide a very rapid means of exploiting air strikes, and of converting their effect – which is usually transitory – to something more permanent.

Air strikes will have a considerable effect on the armoured forces of the enemy, which means that adequate **air cover** acquires some significance for our own tank forces. The tank forces themselves are fairly resistant to air attack. It takes a direct hit or a near-miss to knock out or damage a tank, and the tanks can provide their own air defence through camouflage and anti-aircraft weapons. However tank forces can be hit badly by an air attack if it happens to catch them when they are resting, or their crews are out of the vehicles. Moroever it is difficult to provide protection for the auxiliary arms of the tanks and for the supply train, whose vehicles are mostly soft-skinned. They need special anti-aircraft weapons of their own.

These supporting units also require a complement of **anti-tank weapons** and the latter will also be needed to secure forming-up areas, halts on the march, assembly and rest areas. They can also play an important part in the tank battle by covering the initial points of departure and forward assembly areas of the tank forces, and also their flanks and rear.

In the World War the shortcomings of the **signals and communications systems** greatly impeded the command of tank forces, and their co-operation with the other arms. Tank company commanders were sometimes reduced to accompanying their forces on horseback, to exercise a modicum of control, and they had to make considerable use of runners. Here is the origin of the accusation that tanks are 'deaf'. This shortcoming has now been overcome by that magnificent invention, the wireless telegraph and its relation, the voice radio. Every modern tank has a radio receiver, and every command tank is equipped with both receiver and transmitter.[10] Tank units are now under guaranteed command and control. Inside the larger tanks are various devices to enable the crew members to communicate with one another.

Radio is likewise the principal medium of control between tank units and the other forces, and radios are the main equipment of the signals elements which provide the communications for the tank units and their supporting arms. Telegraphic and optical communications have become generally unusable because of the speed of mechanized forces, the way they have extended in breadth and depth both on the march and in action, and the dust and smoke of the battlefield. However visual signals are used up to company level as a replacement for radios in case of breakdown.

Telephones are employed during quiet spells, when the forces are being held in readiness over long periods of time, and (not excluding the postal telephone network) for approach marches behind our own front line.

Basically the signals elements maintain the communications between commanders and their sub-units, beween commanders and their own superiors, and with whatever neighbouring forces, aircraft and other units are engaged in the common task. The signallers must remain in the closest contact with the commanders to whom they are assigned. In combat these commanders will be right up front with their tanks, which means that armoured radio vehicles with full cross-country capability are essential for the panzer signals elements.

In combat the transmission of **orders** are conveyed to the rapidly moving armoured forces in different and much shorter forms than with the infantry divisions. Reports and orders get through much more quickly when we adopt abbreviated voice procedures and prearranged signals for contingencies. The only way to secure good communications within the tank forces and their co-operation with the other arms is through constant practise, and specialized tactical and technical training. Tank forces are indeed 'deaf' when they do not have signals units of this kind, and the same holds true of their superior officers, their neighbours and their sister arms – deaf to the events which are being played out around them.

Lastly we must devote some words to the question of the **supply** of tank forces and their supporting arms. Until quite recently the objections most frequently raised against a massive programme of mechanization in general, and the setting up of a large tank force in particular, concerned the difficulties in the supplies of fuel and rubber. These objections carried some weight at the time, but we are delighted to be able to say that they will cease to do so in the near future.[11] We owe this to the comprehensive measures that have been taken in the Reich's Four Year Plan to assure the synthetic manufacture of fuel and rubber.

We are still left with the problem of getting the fuel and tyres to the mechanized forces in good time, together with their ammunition, rations, medical services, workshops and replacement personnel. In this respect we must strive to limit the rearward services to the necessary minimum, so as not to complicate the management of the tank forces. The solution lies in the total mechanization of the supply system.

When we trace the process of argument that we have followed over the last few sections, we are left with a number of salient issues which concern the organization and training not just of the tank forces, but of the other arms that are destined to work with them. These issues in fact go to the heart of eternal problems of defence and attack, and the various means by which they might be resolved.

Notes

This entire chapter is derived from an earlier article of the same title: 'Die Panzertruppen und ihr Zusammenwirken mit den anderen Waffen', in: *Militärwissenschaftliche Rundschau*, issue 5/1936, pp. 606-26.

1. The French Army was very short of anti-tank mines in 1940. Alistair Horne, *To Lose A Battle* (Macmillan), 1969, pp. 158 and 165. But mines were later to play a crucial role in armoured warfare, most notably in the Western Desert and on the Eastern Front.

2. This prediction was to hold good for the later stages of the Second World War. One of the reasons for the ease with which the Germans succeeded in 1940 was that at the breakthrough sector they selected on the Meuse the French were desperately short of anti-tank guns. Horne, op. cit. p. 158.

3. In 1937 when Guderian wrote this last sentence it was a bold statement. Today it is one of the most well-worn of military axioms.

4. The concept which Guderian outlines in this paragraph and the last, of attacking the whole depth of the enemy's defences simultaneously, is very reminiscent of the writings of the great Russian military thinker of the inter-war period, Marshal Tukhachevsky, who called the concept 'Deep Battle'. See R. Simpkins, *Deep Battle: The Brainchild of Marshal Tukhachevsky* (Brassey's), 1987, passim.

5. For British reaction to the first tank versus tank clash see Elles to General Staff GHQ, 26 April 1918. B40, Fuller Papers, Tank Museum.

6. In preferring stationary fire to fire on the move Guderian once again demonstrates his grasp of tactical realities. In the 1930s and 1940s this was the only way for tank gunners to achieve accuracy. On the move a tank was a hopelessly unstable gun platform. Yet in Great Britain between the wars the Royal Tank Corps made a shibboleth of fire on the move and this was taught at the RTC gunnery school at Lulworth almost to the exclusion of stationary fire. See Liddell Hart, *The Tanks*, Vol. I, pp. 228-9. Percy Hobart, the leading light of the RTC in the mid-late thirties, the man who raised and trained 7th Armoured Division in Egypt, was a fanatical advocate of fire on the move. See Hobart to Liddell Hart 21 September 1936, para. 4, LH 1/376/35(a)b., Liddell Hart Papers, Liddell Hart Centre for Military Archives (LHCMA) King's College London (KCL). The extent to which this difference in tactical doctrine affected the course of British–German tank battles in the desert is a subject worthy of investigation.

7. In practice the panzer forces were to demonstrate enormous pragmatism and flexibility. In the crucial Meuse crossing operation of 13 May 1940 Guderian's XIX Corps led with its engineer and infantry components, the tanks following later. F. K. Rothburst, *Guderian's XIX Panzer Corps And The Battle of France* (Praeger), 1990, pp. 72-81.

8. Guderian's commonsense and breadth of vision are apparent in the stress he places on the panzer divisions' sapper components. This contrasts markedly with the views of Percy Hobart, the most radical of the British tank enthusiasts, who, at about the time Guderian wrote *Achtung – Panzer!*, was advocating a small and almost all-tank armoured division with no integral sapper component – even though he was an ex-engineer himself. Organization of Higher Mobile Formations, 3 February 1937, LH 15/11/6, Liddell Hart Papers, LHCMA, KCL.

9. Again Guderian's attitude to co-operation between tanks and other arms is worth contrasting with that of Hobart, the leading light of the British Royal Tank Corps at this time. Hobart saw the need for only a tiny amount of infantry (one battalion) in the armoured division and even that was to play little role in mobile operations. LH 15/11/6 and 15/11/7, 28 September 1937, Liddell Hart Papers, LHCMA, KCL.

10. The radio provision in the German armoured divisions was better than in those of any other army at the outbreak of the Second World War. This was an enormous advantage in terms of tactical flexibility. Harris and Toase, *Armoured Warfare* (Batsford), 1990, p. 59.

11. Guderian was rather over-optimistic on the synthetic fuel issue. In fact in the closing stages of the war, especially after they had lost access to Roumanian resources, the Germans suffered serious fuel shortage.

WARFARE AT THE PRESENT DAY

1. THE DEFENSIVE

When the World War ended in 1918, the defensive had attained a strength which was unprecedented for centuries. It was the defensive that had derived the chief benefit from the buildup of the resources of the infantry, artillery and engineers that had taken place during that conflict. The air and tank forces had made the main contribution on the offensive side, but in 1918 they were both arrested in their infancy, and were therefore unable to display their full potential – a fact which exercises a considerable influence even today. Despite the clear warnings of 1918 the inclination is still to underestimate these two new arms of the service, rather than otherwise.

Let us say for the sake of argument that air and tank forces did not exist at the present time, and then go on to examine the implications for the attack and the defence. The unavoidable conclusions are that it would be far more difficult than it was even in 1918 for an attacker to gain a decisive victory over a defender of approximately equal force; that a very considerable superiority in *matériel* and quantity would still offer no guarantee of success, and that if we wished (or for the lack of time were forced) to gain an offensive victory, we would have to cast around for some entirely novel means of attack.

How has the situation on the continent of Europe developed since 1918?

Permanent frontier fortifications have come into being, and they surpass anything seen since the time of the Romans; in some countries they form continuous defensive zones and are in a state of active armament. The garrisons, weapons and ammunition are all securely lodged in shell-proof accommodation; obstacles have been laid out and the facilities for communication have been established. The garrisons are in permanent residence in the fortifications even in time of peace, and they have been separated organizationally from the field army. Skilful use has been made of all the advantages the terrain has to offer, and the natural and artificial defences are complementary. It is fair to assume that, behind the border fortifications, a number of rearward defences already exist, and that others have been laid out for future development. We know from the World War just how rapidly such lines may be strengthened to frustrate attack by greatly superior conventional forces.

Even if we managed to bring off surprise against fortifications of this kind, the penetrations would be checked by fast-moving motorized reserves, giving the defenders time to organize their countermeasures. As early as 1916-18 modern means of transport, and trucks in particular, acquired great and undeniable defensive value.

It is out of the question to think of assaulting fortifications like this with the weapons of 1916; the only result would be that the offensive would be exhausted in a weary battle of attrition in which the attacking forces would find themselves at a severe disadvantage and suffer accordingly.

But it does not end there. We also have to take into account the way that various countries have gone on to organize their defences since 1918. A number of states without permanently mobilized forces have chosen to lay out defences along the most important sectors of their national frontiers. These works will be located in terrain that is inaccessible to tanks, or at least offers protection against all realistically foreseeable kinds of armour. They will have ample anti-tank weapons, enjoying meticulously surveyed fields of fire. All due attention will have been paid to the camouflage and anti-aircraft defence of these assets. Behind ramparts of this kind the defensive will have extraordinary potency against even the modern offensive weapons of aircraft and tanks. Something altogether new in the way of offensive power would be needed to smash such a defence in a reasonably short time.

Nations that have been endowed by Nature with predominantly strong borders, and who are able to defend the rest of their territory in the way just described, possess indeed a high degree of security. If the neighbours of such nations have not followed their example, the fortifications also serve to offer good cover for the deployment of an offensive army.

It is a different matter when countries do not have natural frontiers, and when they are devoid of continuous and strong border fortications on the lines of the Roman *limes* as we find elsewhere. Nations in that situation will be able to confront the attacker only with discontinuous fortresses, helped out at best with light intermediate works. Positions of that kind offer reasonable protection against conventional weapons, but not if the enemy is able to employ air forces and tanks. If the attacker gains surprise, he will not find it particularly difficult to penetrate through the gaps.

When countries are indeed surrounded by a modern Great Wall of China they enjoy such a high degree of security that it seems they could dispense with their tanks, and rely on the strength of their fortifications, the impregnability of their obstacles, and the excellence of their anti-tank weapons. In fact they are very far from doing so. They happen to be just the nations that have created particularly large tank forces, well suited for attacking fortresses, and the ones which subject their armed forces to constant augmentation and modernization. There are two explanations. Either they know that even the strongest fortresses have their Achilles' heel, or they plan to open a surprise attack on their own account – which is perfectly feasible, since they are in a state of constant readiness.

Countries without Chinese Walls must therefore reckon on the attacker gaining initial success with the advantage of surprise, and effecting break-ins of varying speed and depth. For the break-ins the attacker will make little use of his infantry divisions, and still less of his divisions of cavalry; he is much more likely to commit his heavy breakthrough tanks in his first line of battle, with the light armour and all kinds of motorized supporting arms following up. Simultaneously with the ground assault the attacker will strike with his air forces, with the intention of crippling the defenders' air forces, imposing delays on the movement of defending troops – especially the tanks and other motorized units – and destroying the command system. The air and tank assaults will be especially effective if the defenders are slow to set their own forces in movement. If a lack of space compels the defender to do all he can to limit the depth of the break-in, he will have to be particularly quick on his feet, and under an obligation to match the enemy in the air and on the ground with equal, if not locally superior, forces.

As far as tanks are concerned, superiority on at least the local level is attainable only by concentrating all the available forces – scattering the tanks equally among armies, corps and divisions is a recipe for permanent inferiority on the decisive sectors. The business of deciding where to concentrate the armoured forces for the decisive defensive battle will be simplified if, because of difficulties of terrain, the deployment of the larger motorized and tank formations by attacker and defender alike is restricted to particular areas. It would be a grave mistake to commit tanks to areas where you do not wish to stage a decisive battle, or where this is ruled out altogether by topographical obstacles. Weak blocking forces would be adequate for sectors of this kind.

Where do we end up, when we scatter our resources of tanks in a defensive posture, strung out equally all along the front line? We end up by being defeated, just like the British in 1918! In contrast the French held back their tanks for a successful counter-offensive, and this gave them victory in the Battle of Soissons in July 1918.[1]

2. THE OFFENSIVE

Every attacker needs striking power,[2] whether he intends to deliver a strategic surprise, as outlined above, or launch a breakthrough or a counter-blow from a posture of defence.

What exactly do we mean by striking power? Does it reside in our bayonets, in the rifles of our infantry, or even in our machine-guns and artillery? Can they really move quickly enough with the motive capacity of men and horses? Do bodies of riflemen armed with the bayonet and the M 1898 Carbine genuinely represent the striking power of the infantry? Is it realistic to expect that these men, who are effectively defenceless for most of the duration of the combat, can go on to launch a 'storm' against machine-guns, and still show a moral superiority over defenders who are firing from cover? Will we not make the same mistake which in 1806 caused the

Prussians to advance proudly against the foe without firing a shot, and then, so as not to disturb the alignment of their heads, opened battalion volleys without taking aim, and indeed without lying down under enemy fire? The Austrians in 1866, the British in the Boer War in 1899, the Russians in Manchuria in 1904, and the young Germans in Flanders in 1914 – they all relied on the bayonet. And what was the result? Do we really have to go through all that again?

Incredibly enough one is still accounted a heretic if one dares to assail the sacred cow of infantry shock action, as invested in the bayonet. It is timely to repeat what General Field Marshal von Moltke wrote and said on this point more than eighty years ago now: 'Since the defenders have a distinct tactical advantage in a fire fight, and since the Prussian needle gun is better than the infantry weapons of foreign armies, there is more reason than ever for the Prussian army to fight on the defensive.' (*Moltkes taktisch-strategische Aufsätze*, Foreword, xii.) He taught that, 'Even when you are on the attack you must shake the enemy by bringing fire to bear on them, before you carry through with the bayonet.' He cautioned, '. . . in practice this was probably what actually happened in the attack as recommended and employed by Frederick the Great; people at the present day are nevertheless very fond of describing it as "getting stuck in with the bayonet".' (*Moltke*, 56.) He described the action at Hagelsberg in 1813, the great day of the Landwehr with its celebrated bayonet charge, which cost the enemy the grand total of between 30 and 35 dead, and he concluded, 'The statistics indicate that it was not the bayonet attack which determined the outcome at Hagelsberg, but the other way around – the bayonet attack was carried through because the issue had been decided already.' (*Moltke*, 57.)

In the era of machine-guns and hand-grenades the bayonet has surrendered still more of its significance. As early as 1914 striking power resided in firepower – which for the infantry meant their machine-guns and other heavy weapons, and for the higher, divisional level, the artillery. If this striking power was adequate, as it was on the Eastern Front, and in Romania, Serbia and Italy, the attacks succeeded. Where it was inadequate, as on the Western Front, the attacks failed.

Striking power as expressed in firepower attained gigantic proportions in the World War, whether measured in quantity of ammunition, calibre of artillery or the length of the bombardments. And yet as a general rule it was incapable of smashing the enemy resistance quickly enough or completely enough to attain more than a deep break-in of the defensive system; or at least not on the Western Front, which was the decisive theatre of the war. On the contrary, the long duration of the bombardment, which was considered necessary to gain the decisive effect, gave the defenders time for countermeasures – bringing up their reserves, or if necessary falling back. Very often the first sign of an imminent offensive was enough to trigger a decision on the part of the defenders to withdraw; such counter-moves were prepared with care, so that the enemy blow fell on empty air at the decisive moment, or the attacker simply abandoned the assault altogether. The best examples are the German withdrawal to the Hindenburg Line in 1917, and the French fall-back at Reims in 1918.

The World War proved that striking power does not consist in firepower alone, however furious and long-drawn-out it might be. It is no good converting firm ground to a lunar landscape by an unaimed area bombardment; you must bring fire to bear on the enemy by closing to close range, identifying the targets that pose the greatest hindrance to the attack, and annihilating them by direct fire.

In the time of Frederick the Great it was still possible to come at the enemy with cold steel in the form of the infantry bayonet and the cavalry sword, relying on the muscle power of men and horses. Those days have long passed, and even in the Seven Years War General von Winterfeldt could write to the king: 'We simply won't make it if we advance with shouldered muskets and without firing.' In fact a precondition for the success of shock action was that the enemy were already rattled by fire. Even the celebrated charges of the Bayreuth Dragoons at Hohenfriedberg [1745] and Seydlitz's at Rossbach [1757] were aimed at infantry that had already been shattered. Attacks against unbroken infantry were indecisive, as was shown at the Battle of Zorndorf [1758].

The need for preparatory fire before an attack grew in proportion to the increasing range, speed of fire and penetrating power of weapons. These developments chiefly benefited the defensive, and they culminated in the attritional or artillery battles of the World War. And yet at the present day even the strongest firepower is no longer adequate to permit us to advance with sufficient speed and 'bring fire to bear on the enemy'. The only weapon of any use in this respect happens to be an ancient one, by which we mean armour. Armour had fallen out of use in olden times not because it could not be made thick enough to offer protection against firearms, but because men and horses were simply not strong enough to carry armour of such dimensions! This power was then supplied by the invention of the internal combustion engine. It was now possible to move armoured vehicles and their crews unscathed by small-arms fire until they closed with the enemy, and were able to bring them under direct fire and wipe them out. Motorized armoured vehicles also had the crushing capacity to cross and destroy the dreaded belts of barbed wire, and the obstacle-crossing capability to overcome trenches and other obstructions. In late 1917 and in 1918 the true striking power of the Allied armies was therefore inherent in the tanks, after the 'impregnable' Hindenburg Line had been broken through at Cambrai in a single morning.

What after all is striking power? It is the power that enables combatants to get close enough to destroy the enemy with their weapons. Only forces that possess this capacity can be reckoned to have true striking power, in other words, have true offensive capacity. After the experiences of the last war we are not boasting when we proclaim that, out of all the weapons of ground warfare, the tank has this striking power to the highest degree. Post-war developments have not presented the avid military world with anything better. For better or worse soldiers will have to come to terms with the tank, however difficult it might be for individuals to change their ways.

When the greatest striking power of the offensive resides in a particular weapon, this weapon must claim the right to use that power according to its own rules. It will be the main battle-winning weapon, wherever it is put into action, and the other weapons must accommodate themselves to its needs. What is at issue is therefore not helping a particular weapon, however tradition-laden it might be, to score a measure of success, but of winning the battles of the future, and of winning them so completely, so speedily, and so comprehensively that they will bring the war to a rapid conclusion. All the weapons must co-operate to this end, gauging their own performance and demands according to those of the arms that have the greatest striking power.

The tank forces are at once the youngest of all the arms, and the one with the highest degree of such striking power. They must therefore advance their claims on their own account, since there is no country in the world where the other arms will concede these of their own free will. The more effective the developments in anti-tank defence, the more difficult will be the armoured attack, and the more forcibly and loudly the tankmen must press their demands.

Now, as before, these tactical requirements come down to three: **Surprise**; **Deployment *en masse***; **Suitable terrain**.

These are the preconditions for the success of any kind of tank attack, and they will determine the organization of the armoured forces in war and peace, their weapons and development, and finally the selection of their commanders and troops.

Surprise may be attained through speedy and well-concealed movements, through the appropriate preparation and execution of the attack, and through new weapons of unprecedented capability. The rapid execution of the armoured attack is of decisive importance for the outcome of the battle; the supporting arms that are destined for permanent co-operation with the tanks must accordingly be just as fast-moving as the tanks themselves, and they must also be united with the tanks in an all-arms formation in peacetime. It is a different story when the tanks do not have supporting arms of this kind. They end up going into action with forces that have never settled into teams with them, that are slow-moving, and which therefore deprive the tanks of the capability to thrust fast and deep into the enemy. In other words they throw away their greatest asset.

Great things may hang upon military equipment of a novel kind – impenetrable armour, for example, or an excellent gun, or an unusual turn of speed. When preparations are proceeding in peacetime it is therefore most important to preserve the greatest secrecy in the field of military technology. As already mentioned, the prime example of security and its benefits is represented by 'Big Bertha', the 420mm gun which shattered the Belgian and north French fortresses in 1914.

The principle of **deployment *en masse*** – the concentration of forces where we seek to gain the decision – is actually one that is valid for all arms. And yet in Germany and elsewhere there are many voices who maintain that the contrary is true of tanks. This is a violation of one of the first principles of war, and we cannot accept it passively in peacetime if we are to avoid the most condign punishment in the case of hostilities. If we once accept the principle of deployment *en masse* – the concentration of force on the decisive point – we must draw the necessary conclusions in terms of organization. Deployment *en masse* can be accomplished in actual warfare only if the tank forces and their commanders have learned how to fight in large formations in peacetime. In the case of mobile troops, and their leaders, it is very much more difficult to improvise from the ground upwards than it is with the infantry.

As for the **terrain**, the tank forces should be committed only where there are no obstacles that exceed the capacity of their machines; otherwise the armoured attack will break on the terrain. It is, for example, thoroughly misleading on exercise to dig a ditch that is impossible for a particular type of tank to cross, then force the tank to tackle it by binding orders, and finally broadcast the 'failure' of the machine or the tank forces as a whole. It is no less absurd to require light machine-gun tanks to assault a fortress or a large city. This would not even be demanded of light artillery – a big task calls for a heavy calibre of weapon. Tanks have a certain capacity, just like men and animals; when one's demands exceed that capacity, they will fail.

Since we cannot hope to find favourable terrain for tanks everywhere, we must strive to employ them where they can move and show their striking power to the best advantage – or at least in sufficient force, breadth and depth, and with the element of surprise. It is also a question of organizing the tanks in mixed armoured formations that are capable of meeting their assigned tasks, and of training their commanders in an appropriate way.[3]

In the last war the tanks invariably came off badly when they were fed into action in penny packets – and that was at a time when the Germans had hardly any kind of organized defence. In a future war, however, both sides will have had the opportunity to come to terms with tanks and devise appropriate measures of anti-tank defence in time of peace. In such a context there will be horrible consequences attendant on any misuse of tanks. How might such a misuse arise? It is most likely to derive from a mistaken appreciation of the respective effectiveness of the defensive and the armoured offensive. A defective organization of the armour could all too easily result.

As far as ground combat is concerned, we consider that the best chance of success for the offensive in modern warfare lies in deploying armour *en masse*, in suitable terrain and with the advantage of surprise. We must emphasize that any success on the part of the attack must be exploited speedily by the other arms, if it is not to lose its effect after a short period of time. But it is also our conviction that aerial warfare, as well as ground combat, must be influenced to a significant degree by the existence of tank forces.

3. Aircraft and Tanks

We have often mentioned the role of aerial reconnaissance and tactical aircraft in the support of armoured attacks. But it can also work the other way round, with the operations of tank forces promoting the ends of aerial warfare. One could imagine how at the beginning of a war the armoured forces could strike at vital enemy airfields or other relevant objectives close to the border; again, after successes on the ground at a later stage of the war, the tactical aircraft, airlanding troops and tank forces could be assigned common objectives deep in the enemy rear, with the aim of breaking the enemy's power of resistance with the least loss of life. This is a concept of warfare which has so far received little attention – the pundits have been too preoccupied with the questions of infantry support and the initial tactical decision in ground combat. But we do not have to be out-and-out disciples of Douhet[4] to be persuaded of the great significance of air forces for a future war, and to go on from there to explore how success in the air could be exploited for ground warfare, which would in turn consolidate the aerial victory. Here again it comes down to striving for a common victory, and looking beyond the interests of an individual arm of the service.

4. Questions of Supply and Transport

The extensive mechanization of the army has raised two important issues: how will the army as a whole be supplied with fuel, spares and replacement vehicles? And how will we be able to move our large mechanized formations, and especially the ones that are road-bound? A positive answer to these two requirements is a precondition for the deployment of large tank formations, not least for their use in the operational dimension.

As far as **fuel** is concerned, German's consumption in 1935 came to 1,920,000 metric tons. The figures for 1936 are made up of: 1,382,620 tons of imported petrol and oil; 444,600 (approx.) tons of synthetic fuel; 210,000 tons of spirits (*Spiritusbeimischung*).

These means that in 1936 two-thirds of German peacetime needs still had to be met by imports. However the Four Year Plan has provided in a comprehensive way for synthetic fuel production, which will change the picture significantly and free us in the foreseeable future from the need to import petrol and oil.

In addition the calls on fuel can be diminished by using alternative sources of energy, which will have their primary application in civilian life; here we have in mind various forms of motors driven by gas or electricity.

It will likewise not be long before Germany is independent of imports of foreign rubber.

An efficient motor and engineering industry is a prerequisite for a **sustained replenishment of military vehicles and spares**. The following figures will give an overview of how we stand in comparison with the major industrialized nations:

Production of Motor Vehicles

1935		1936	
United States	74.1%	United States	77.2%
Britain	9.1%	Britain	7.8%
France	5.3%	Germany	4.8%
Germany	4.7%	France	3.5%
Canada	3.1%	Canada	3.4%
Italy	1.2%	Italy	0.9%
Others	2.5%	Others	2.4%

The United States, Canada and Germany have therefore been able to increase their share of world production, with Germany rising from fourth to third place – a favourable position which means that in the event of war we will be able to maintain the level of our mechanized forces and our motorized rearward services.

In all of this it is important to site most of our centres of production in localities where they are secure, and out of the way of immediate ground or air attack. The products must also be assigned in a rational way to the various end-users – the army, the navy, the Luftwaffe and the domestic economy. Moreover the productive capacity of the factories must be guaranteed by ensuring that they can hang on to their skilled labour and their engineers and fitters in the event of war.

The **network of highways and roads** is of fundamental importance for the movement of mechanized forces, and especially at the outbreak of war, when large numbers of commercial road-bound vehicles will have to be requisitioned and incorporated in the field formations. The construction of German highways has suffered from many decades of parsimonious neglect. The reason was that the German federal government, as the responsible organ, devoted most of its attention to building railways, and dumped the responsibility for the roads on the lesser authorities – the provinces, circles and local administrations – which meant that hundreds, indeed thousands of bodies were responsible for maintaining the roads. The advent of motorized traffic had no immediate effect on this traditional structure, and the 'autonomous rights' of the provinces remained sacrosanct, even when they had become largely a dead letter.

The Führer is a man of vision. He has recognized the enormous significance for motorized traffic of a programme of road construction which is massive in scale and carried through on consistent principles. The Reich has taken under its wing the main through roads, and instituted the construction of highways of a unique kind, our autobahns. The initial programme is for 7,000 kilometres of autobahn which will link the main cities of the country. These highways are broad and unencumbered by crossings and counterflowing traffic, and they will permit the maintenance of high and steady speeds over long stretches – permit us, in other words, to exploit powered vehicles to the full for the first time ever.

The military significance of good roads for powered vehicles is self-evident. But a peacetime road network, no matter how dense, can never meet all the unexpected tactical and operational demands of warfare. In the old days soldiers usually had to make do with the roads available in peacetime, which were built with economic considerations in mind. But the War of 1914-18 demonstrated demands on road construction of a massive order; we just have to call to mind the road conditions in front of Verdun, on the Somme and in Flanders, the endless plank roads on the Eastern Front, and the difficulties which the armies experienced with the roads in Mesopotamia and Palestine.

The achievements of the Italians in building roads in Abyssinia are particularly impressive, and generally it was only the existence of these roads that permitted them to make extensive use of their motorized forces.

In this context we may draw the following conclusions from the Italian campaign against Abyssinia:

1. The peacetime road network has an undeniable influence on the operations of an army and how it works at the tactical level; both sides usually avail themselves of the network, which is known to them both and set down on maps.
2. The peacetime road network is, however, not incapable of expansion; in wartime it can and must be adapted to the operations we have in mind, and extended accordingly.
3. Some extensions may be in the form of permanent roads – but these are expensive in terms of time and labour, and they will be located by the enemy aerial reconnaissance. It will often be sufficient to construct soft-topped routes which are good enough for tracked and cross-country vehicles. Roads of this description can be built at some speed, and under favourable conditions they may escape undetected by reconnaissance for a considerable time.
4. The speed with which soft-topped roads can be built is something that helps the movement of the other arms, as well as the sudden irruption of mechanized forces.
5. The mobile armies of the future must therefore have a sufficient number of road-construction units, equipped with modern machines and tools.

5. THE MOST RECENT EXPERIENCE OF WAR

The most recent examples of the deployment of tank forces are represented by the Italian campaign in Abyssinia, which we have just mentioned, and the fighting which is still in progress in Spain.

In Abyssinia the Italians put into the field about 300 Fiat Ansaldo tanks. They were equipped with machine-guns only, and did not have traversible turrets. The fixed positions of the machine-guns put the Italians at a disadvantage, and particularly when the tanks were employed one at a time, which enabled the natives to board the machines and kill the crews through the vision slits, which were inadequately protected. On the other hand the tanks operated to generally good effect in spite of the difficulties presented by terrain and climate – neither the sandy deserts nor the high mountains proved to be insurmountable obstacles. However there is a limit to the lessons that are relevant for warfare in Europe, since the Abyssinians had no anti-tank defence and no armour of their own.

When we look at the categories of the Italian tanks, we see that they performed their tasks well, whether the armoured reconnaissance vehicles on scouting missions, or the tanks which operated with motorized infantry in a variety of assaults. Altogether the armour helped the Italians to finish off the campaign as quickly as they did.

The war in Spain offers a somewhat greater range of experiences. As far as we can tell, the Reds use the Russian-made Vickers 6-ton tank, which is armed with a 4mm gun and one or two machine-guns. With its full equipment the tank weighs eight tons, and its most important areas are impervious to armour-piercing small-arms fire. The Nationalists have employed only machine-gun tanks, which are armed with two machine-guns mounted in a revolving turret. These too are proof against machine-gun fire. Apart from some captured vehicles, it appears that General Franco's forces have no gun tanks proper, though they are equipped with a number of 37mm anti-tank guns.

So far no more than fifty tanks have appeared in action at a single time, which leads us to suspect that neither side has a particularly great number of tanks, or any heavy tanks with thick armour and powerful guns at all, in addition to the light tanks already mentioned. When we look at the numbers of machines available, and their types, there seems little prospect of tanks being able to bring off a rapid and decisive victory in this conflict.

It is not at all surprising that the Nationalists have held back from sending their tanks across open country in the face of the Russian tank guns' long-range fire; the enemy show the same bashfulness in respect of Franco's anti-tank guns.

The terrain on the western sector of the front must be accounted difficult in the extreme.

Of the three conditions for a successful tank battle – surprise, deployment *en masse*, and suitable terrain – only the first is at all attainable, and then only if the belligerents are skilful enough. So far neither of the two parties seems to have attempted to deploy all its available forces in one go; as for the choice of terrain, we can only comment that a great city like Madrid is not exactly a suitable setting for an attack by machine-gun tanks.

The reports of the fighting are fragmentary, but it does seem that tanks have been committed in all the major actions, and that as a general rule the infantry do not advance until the armour has done its work. In the process the tanks have inevitably suffered a number of losses, but this a fate which they share with all arms.

On the technical side there will undoubtedly be an accumulation of lessons over the course of time, even if it is premature to assess them just now.

As far as the tank crews and commanders are concerned, the war has emphasized the old need for long service and professional training – a few weeks are simply not enough to enable the Spanish soldier to acquire a complete command of the modern machines of war. There are also indications that the high command still does not have the experience to be able to use tanks properly.

With this we have reached the limits of what we can deduce in the way of conclusions and lessons from the available reports of what has been happening in Spain.

In our opinion neither the war in Abyssinia nor the Spanish Civil War[5] can offer a kind of 'general test' of the effectiveness of armour – the numbers and models of the tanks are too few, and the actions have been too one-sided and too small in scale. However the two conflicts may well provide a number of pointers as to the technical and tactical development of armour, and we will have to investigate them attentively and derive what lessons we can. All the same they have in general terms given us no reason to abandon the principles we have already set out.

Notes

1. From a post-Second World War vantage-point this passage has a certain irony. The French, whom Guderian here praises for their concentration of armour for the counter-stroke at Soissons in 1918, in 1940 made precisely the error which Guderian accuses the British of making in 1918, spreading their armour right across the front. Icks, *Famous Tank Battles* (Profile), 1972, p. 103.

2. The German word Guderian uses (here translated as striking power) is *Stosskraft*. Kenneth Macksey has rendered it as 'dynamic punch'. See Macksey, *Guderian* (Macdonald and Jane's), 1975, p. 45. There is in fact no exact English equivalent.

3. The basic principles of large numbers, suitable ground and surprise had been laid down by Swinton as early as February 1916. See Notes On The Employment Of 'Tanks' by Swinton, Stern Papers, LHCMA, KCL. Printed in Swinton, *Eyewitness* (Hodder and Stoughton), 1932, pp. 198-214. Guderian adds the critical insight that tanks in the future will only fulfil their potential when fully integrated with other arms in a well-balanced mechanized formation under commanders trained to think in all-arms terms. It was this last insight which (generally speaking) the British Royal Tank Corps, and later the Royal Armoured Corps, did not adequately share in the late thirties and early forties. Of the British thinkers, George Lindsay was the nearest to seeing the point but his influence tragically waned after 1934. See Winton, *To Change An Army* (Brassey's) 1988, pp. 177-83.

4. Douhet was an Italian theorist who, in the 1920s, preached in his book *The Command of the Air* (1928) the doctrine that the next war would be decided by massive bombardment of cities at the opening of hostilities. Guderian believed in air power but is here clearly rejecting Douhet in favour of the closest possible co-operation between air forces and mechanized ground forces. Guderian had explored the impact of air power on ground operations in previous writings. See 'Truppen auf Kraftwagen und Fliegerabwehr', in *Militär-Wochenblatt*, Nr. 12, 25 September 1924 and 'Der Einfluss der Luftwaffe auf die Infanterietaktik' in the Austrian journal *Militärwissenschaftliche und Technische Mitteilungen*, LIX, 1928, pp. 507-12.

5. Guderian was right virtually to dismiss Spain as a source of tactical and operational lessons for armour. Field Marshal Sir Cyril Deverell, the British Chief of the Imperial General Staff at the time, similarly dismissed the Spanish experience on the grounds that tanks had been too badly handled there to draw any lessons. He continued to believe that tanks were very important for the breakthrough role. Liddell Hart, however, was increasingly doubtful whether tanks could overcome the great strength of the defence and Spain appears to have reinforced these doubts. Talk with Deverell, 29 June 1937, LH 11/1937/56, Liddell Hart Papers, LHCMA, KCL.

CONCLUSION

Little more than twenty years have passed since tanks first appeared on the bloody Somme battlefield; this is a short period in the span of history. But modern technical development has acquired storm force – it carries the economy along with it, and it has speeded interchange among mankind. The whole existence of the nations has been set into a more lively motion.

It would therefore be mistaken to confine ourselves to purely technical considerations, when there are far more wide-ranging issues at stake.

The World War was unleashed by a turmoil of economic and social circumstances, and the most advanced nations of the world were plunged into the whirlpool. There were many who looked forward to some kind of improvement in mankind and the nations after the war, but they were disappointed. On the contrary, there is good reason to fear that the ideological, political and religious contradictions have become as acute as the economic ones. We have no means of telling where the path will lead. But we must recognize that only strong nations can survive over the long term; desire for self-preservation can be turned into reality only when it has the necessary power behind it.

The consolidation of German power is the task of the politicians, technology, the economy and the Wehrmacht.

The stronger the Wehrmacht is in terms of arms, equipment and the spirit of command, the more secure will be the maintenance of the peace. The tank is the most modern weapon of ground warfare, and we hope that a description of how it came into being will provide some insight into its future development.

On many issues there still exist differences of opinion of a sometimes quite fundamental nature. Only time will tell who is right. But it is incontrovertible that as a general rule new weapons call for new ways of fighting, and for the appropriate tactical and organizational forms. You should not pour new wine into old vessels.

Actions speak louder than words. In the days to come the Goddess of Victory will bestow her laurels only on those who are prepared to act with daring.

BIBLIOGRAPHY

Bruchmüller, Georg. *Die deutsche Artillerie in den Durchbruchsschlachten des Weltkrieges*, Berlin.
de Gaulle, Charles. *Vers l'armée de métier*, Paris.
Dutil, *Les Chars d'Assaut*, Paris.
Eimannsberger, L. von. *Der Kampfwagenkrieg*, Munich.
Fuller, J. F. C. *Erinnerungen eines freimütigen Soldaten.*
Hanslian, Dr Rudolf. *Der chemische Krieg*, Berlin.
Heigl, *Die schweren französischen Tanks. Die italienischen Tanks.*
– *Taschenbuch der Tanks*, Munich.
Kurtzinski, M. J. *Taktik schneller Verbande*, Potsdam.
Ludendorff, Erich. *Meine Kriegserinnerungen*, Berlin.
Martel, Giffard, *In the Wake of the Tank. (Im Kielwasser des Kampfwagens)*, Berlin.
Moltke, Helmuth Graf von. *Militärische Werke. Taktisch-strategische Aufsätze*
Oskar, Prince of Prussia. *Die Winterschlacht in der Champagne*, Oldenburg.
Poseck, Maximilian von. *Die deutsche Kavallerie 1914 in Belgien und Frankreich*, Berlin.
Santen, Hermann von. *Die Champagne-Herbstschlacht*, Munich and Leipzig.
– *Schlachten des Weltkrieges*, vols. XXXI, XXXV, XXXVI, Oldenburg.
Schwertfeger, Bernhard. *Das Weltkriegsende*, Potsdam and Berlin.
Swinton, Sir Ernest Dunlop. *Eyewitness*, London.
The Encyclopaedia Britannica, London.
Reichsarchiv, *Der Weltkrieg 1914-1918*, vols. I, V, VI, VII, IX, X.
French official history: *Les Armées Françaises dans la Grande Guerre, vols. I, II, V, Paris*
British official history: History of the Great War.
Military Operations, *vol. II, London.*
Belgian official history: La Campagne de l'Armée d'après les documents officiels, Brussels.
France militaire, Paris.
Militärwissenschaftliche Rundschau, 1936-7, Potsdam.
Militär-Wochenblatt, 1934-6, Berlin.
Revue d'Infanterie, Jan-Feb 1932; Apr-Dec 1936, Paris.

Revue des deux Mondes, Paris.
Versailler Vertrag. Reichsgesetz-Blatt, 1919.
Vierteljahreshefte fär Truppenführung und Heereskunde, 1910-13.

EDITORS' ACKNOWLEDGMENTS

The editors received help from a number of people in the preparation of this edition of *Achtung—Panzer!* In particular they would like to thank Professor Wilhelm Deist and Hauptmann W. Heinemann of the German Military Historical Institute at Freiburg who provided copies of Guderian's previous publications in the German military press; Mr. Keith Simpson who read Paul Harris's introduction in draft and made some useful comments; and Mr. Andrew Orgill and the staff of the library of the Royal Military Academy Sandhurst who facilitated the project in several respects.

INDEX